The Updated HVAC for Beginners 2024

[5 in 1] The Simplified DIY Guide + VIDEO COURSE to Heating, Ventilation, and Air Conditioning Systems | Step-by-Step Procedures & Practical HVAC Tips & Tricks

[Bonus: Video course + Audiobook]

BRANDON DENNEY

Printed in USA

First Edition: Jan, 2024

Your AUDIOBOOK is HERE!!!

Thank you once again for purchasing my book.

This is a thank you gift from me to you as an appreciation for being part of my journey.

To get the Audiobook kindly type in this link on your browser:

http://tinyurl.com/32xps4un

Or

Scan the QR code below for quick access:

Table Of Content

Introduction

Nestled between gentle slopes and evergreen trees, the sleepy community of Crestwood was home to a young man by the name of Alex Turner. Alex had a strong desire to learn about the complexities of the world around him and was an aspirational person. His modest but comfortable house, filled with bookshelves and displaying everything from modern technology to old literature, was a tribute to his passion for self-improvement. But there was one area that Alex had spent far too much time avoiding: the mysterious world of HVAC (heating, ventilation, and air conditioning) systems.

As the changing seasons swept over Crestwood, Alex struggled with the intense heat. He would curl up beneath blankets on the cold winter evenings, and his little house would become a sauna on the hot summer days. As frustration began to seep into his life, he set out on an enlightenment quest that would permanently alter his perspective on temperature management and comfort.

When golden leaves decorated the trees and the smell of pumpkin spice filled the air one crisp autumn morning, Alex stumbled across a worn book in the dusty corner of his neighborhood bookshop. He was drawn to the title, which promised ease of understanding and mastery: "The Updated HVAC for Beginners 2024: [5 in 1] The Easy Do It Yourself Handbook + Video Course on Heating, Cooling, and Fans | Comprehensive Step-by-Step Instructions, Useful HVAC Tip & Tricks, and Additional Resources."

Alex turned the pages with great anticipation, intrigued by the prospect of taking back control of his living circumstances and uncovering a universe hidden behind cryptic terms and arcane symbols. Renowned HVAC specialist Brandon Denney wrote the book, which drew Alex into the center of the HVAC world with an engrossing story from the beginning.

The story of invention and development told by Brandon Denney's writing traces the origins of HVAC systems to prehistoric societies that used clever strategies to control environmental factors. Alex was enthralled by the clever ways our predecessors used nature to create cozy living places, from the Roman hypocausts to the elaborate windcatchers of Persia.

The story moved smoothly to the technical wonders of the contemporary HVAC period as he turned the pages. With a passion that came through in his remarks, Brandon Denney examined the complex dance of HVAC (heating, ventilation, and air conditioning), each component a crucial part of the comfort symphony that reverberated through homes and offices alike.

The book was a knowledge gold mine that included practical applications in addition to theoretical underpinnings that could empower even the most unskilled do-it-yourself enthusiast. Its pages included detailed instructions that shone a guiding light-like clarity on the once-difficult path of HVAC maintenance. It was more than just a guidebook; it was an odyssey, and Alex was prepared to set off on his life-changing adventure.

With the addition of a VIDEO COURSE, Brandon Denney's instruction went beyond the written word and into a new realm. Alex pushed play on the first video, and the screen was animated with explanations that helped make sense of the intricate parts of HVAC systems. Brandon seemed to be physically there in the space, directing him through the maze-like arrangement of compressors, pipes, and ducts.

As each chapter went by, Alex's self-assurance increased. He gained knowledge on how to solve typical HVAC problems, interpret mysterious error codes, and maximize the effectiveness of his system. The story developed like a gripping thriller, providing useful advice and pointers that turned him from a bewildered newcomer into a self-assured HVAC expert.

Brandon, nevertheless, didn't stop there. The book revealed cutting-edge methods that made it harder to distinguish between amateur and professional work, giving Alex more insight into and authority over his living space. He became a zoning expert, adjusting the temperature in every area to an exact science, and he discovered a world of smart thermostats that moved in time with his way of living.

Alex was overcome with a profound feeling of thankfulness as he took in the abundance of knowledge. With its captivating story and thorough methodology, the book not only gave him the tools to overcome his HVAC difficulties but also kindled a fresh interest in comprehending the complex relationship between climate and humankind.

More than just a manual, "The Updated HVAC for Beginners 2024" was a traveling companion on Alex's path to expertise. Motivated by a desire to give back, he imagined a society of do-it-yourselfers, each one equipped with the skills found in those pages. Alex went out to tell his tale, full of gratitude and brimming with newfound understanding, knowing that HVAC mastery had the potential to change the lives of many people.

Equipped with the knowledge found in this book, Alex Turner led the way as a new age of Crestwood's seasons unfolded. The warmth of prosperity emanated from every nook and cranny of his immaculately maintained house, and the cool wind of achievement surrounded him. The once mysterious HVAC industry had turned into his playground, and his hands had created a comforting melody that reverberated throughout his existence.

BOOK I: THE FUNDAMENTALS OF HVAC

Introduction

Our journey starts with a nod to the past and contemplation of the age-old dance that people and the elements performed together. The origins of temperature management are examined in "An Overview of HVAC History," where we look at the primitive methods of heating and cooling that served as the prototype for today's technological wonders. This chapter traces the history of comfort, from the flickering flames of the first fires to the invention of complex HVAC systems. It is a story that crosses ages and cultures.

Chapter 1: An Overview of HVAC History

The efficiency of the HVAC systems in use before was not as high as it is now. Since then, a lot has changed. We will now examine the methods that people in the past used to provide ventilation, heating, and cooling in their dwellings.

Central heating was first used by Greek civilizations to heat their dwellings. They constructed a basement in their houses to accomplish this to maintain a fire. Since hot air rises, the heat produced in the basement would permeate the whole house. To improve air circulation, homes with high ceilings and plenty of windows were constructed in the past. As a result, ductwork was created and utilized to transport air throughout the house. We still employ ductwork in our contemporary systems because of its enduring trait.

The Chinese are the source of an early AC system example. The Chinese found that applying flowing air to the skin produces a brief cooling sensation. Thus, manually operated fans were created. Early Greeks and Romans covered storage holes with snow and tree branches to keep food cold. The storage pits function as the earth's natural cooling system. Consequently, food was stored and didn't rot for a few days.

Chapter 2: Components and Their Operational Functions

Do the parts of the HVAC system need to be known to us? HVAC systems are designed to do more than simply regulate the temperature of a space—mostly indoor spaces—they also must provide clean air to create a comfortable atmosphere when required. As a result, it would be quite helpful to understand the fundamentals of their parts to both appreciate the technical talent that went into them and to make it simpler to determine what components your system could need to function as intended.

HVAC refers to the system that provides heat and air conditioning for your home. You can count on the furnace part of your HVAC system to keep you cool and comfortable in the summer and warm in the winter. Even though some houses are heated and cooled by boilers, radiant floor heating, heat pumps, ductless mini splits, or window air conditioners, let's examine the eight essential parts of a home HVAC system, which includes the classic furnace and split-system air conditioner combination.

What Constitutes an HVAC System's Main Components?

The list of HVAC parts we will cover in this book is as follows:

1. Thermostat
2. Heat generator
3. Heat exchanger
4. Blower
5. Condenser Coil or Compressor
6. Evaporator coil
7. Air Ducts and Vents

HVAC System Components Diagram

Almost every HVAC system has certain key components, which are shown in the accompanying diagram. While there are variations between various HVAC system setups, the fundamental idea and HVAC parts are essentially the same in all of them. The functioning cycle of an HVAC system is shown in the accompanying figure, which may be used for both heating and cooling purposes by varying the locations of heat absorption and rejection.

HVAC stands for heating, ventilation, and air conditioning; as a result, air cleaning and ventilation procedures are often included in the design of these systems.

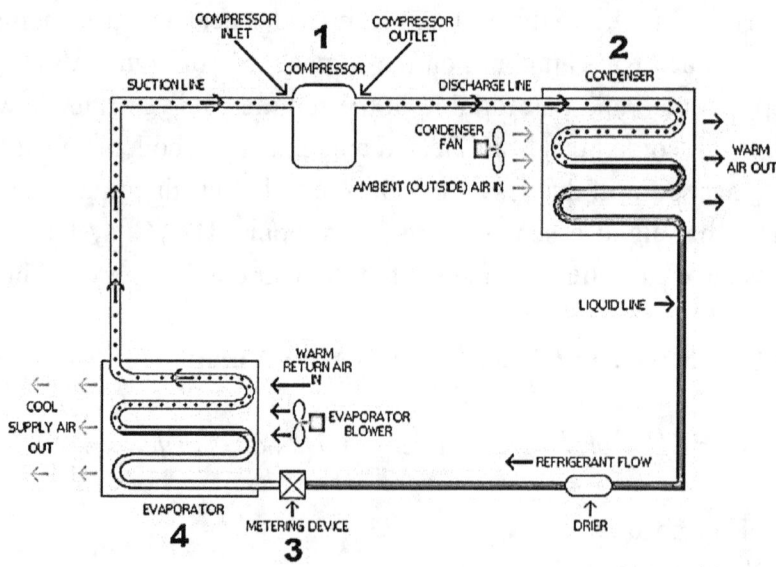

Parts of HVAC System

We shall go more into the main HVAC system components in the next sections.

Thermostat

A thermostat's temperature sensor determines when to switch on or off the air conditioner or heater. There could be many thermally regulated zones, and a thermostat is needed for each one. The thermostat has to be placed as far away from regions where there is a significant temperature differential from the target space's mean temperature.

Heat Generator

One of the most important parts of the HVAC system for heating is the heat generator. In these devices, heat is produced, for example, by extracting fuel energy inside a combustion chamber, also known as a furnace. The air or another fluid, such as water, will then be heated by hot flue gasses, which will subsequently heat the air when it enters the air-conditioned space. The air conditioning might also be heated by electric heat generating.

Furnaces are the most popular kind of heat generator, however, there may be other options as well. For these HVAC system components, it is crucial to take resource management and pollution emission into account as well as combustion efficiency.

The correct and full interaction of fuel with oxygen within the furnace ensures that no fuel is wasted, which is known as combustion efficiency. The furnace's efficiency might also be increased to ensure that the heat is transferred as efficiently as possible—with the least amount of loss possible—to the next medium, which could be water or conditioning air. In general, the following factors should be taken into account: ideal heat transmission, well-shaped glow sticks or igniters, safe operation, and enough fuel and air mixing within the furnace.

Heat Exchanger

One of the parts of the HVAC system that transfers heat from the heat production unit to another fluid is the heat exchanger. When necessary, certain control systems will turn on the furnace or electric heating components to adjust the temperature of the air going through the heat exchanger.

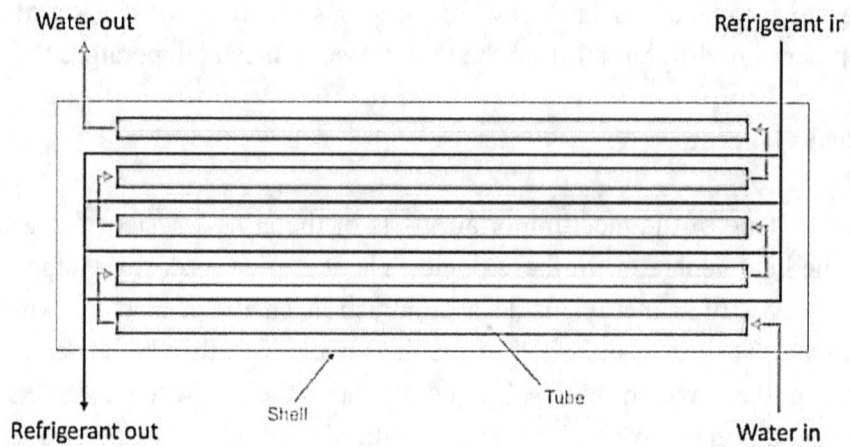

To heat the desired room, heat is often transmitted straight to the chilly air. In this instance, heat absorption is used to transmit energy to the air by forcing air through hot flue gas tubes or electric heating components.

Certain safety precautions must be taken since the majority of heat generators use fuel as their energy source. The reason for this is that to reduce combustion temperature and hence emit less NOx, combustion systems often run with surplus air. Thus, one of the reaction's results would be carbon monoxide. Therefore, carbon dioxide leakage into the air going via the flue gas tubes presents a safety risk for heat exchangers. CO is an odorless, colorless gas that, in high concentrations, may cause headaches, nausea, dizziness, and even death. As a result, detectors must be set up to keep an eye out for these leaks.

Blower

One of the HVAC system's components, the blower, forces air through the heat exchanger and into the air ducting, which transports the heated air to its destination. Through a shaft, an electric motor powers the blower. The motor speed may be changed to alter the airflow. These motors need to be kind with variable speed.

When fewer air volumes are needed, variable-speed motor blowers will gradually increase in speed and produce less noise. These blowers would have reduced operating and maintenance expenses because of the steady speed increase which would also lessen the unit's energy consumption and reduce wear and tear on the spinning components.

Condenser Coil or Compressor

The compressor, often called the condenser coil, is typically housed outdoors and is one of the key parts of the HVAC system. The heated refrigerant gas is sent to the compressor, which releases heat into the surrounding air and transforms it into liquid. The evaporator coil receives this liquid refrigerant via copper or aluminum tubes after that. A fan will speed up the condensation process by increasing the quantity of airflow across the coils.

Evaporator Coil

One interior component of the HVAC system that gets the condensed refrigerant liquid from the compressor is the evaporator coil. By using spray nozzles to improve the rate of refrigerant evaporation upon contact with the warm air in the room, the liquid refrigerant is atomized.

Warm air from the room is forced via return ducts and onto the evaporator by fans. The heated air uses the atomized refrigerant to reject heat and cools down before being redirected via the ducting back into the rooms. The moist air condenses on the cool evaporator coil, lowering the air's moisture content as it passes over it. The air seems much colder due to the reduction in humidity, which increases the effectiveness of the cooling process. The cycle would then be repeated by transferring the heated gas back to the condenser coil.

Air Ducts and Vents

Ducts are used to transport air to various HVAC system components. To provide the zone with high-quality air, proper ducting is necessary. There may be noise when the system is operating due to duct leakage. Furthermore, bad air ducting might cause an airborne stench and an overabundance of moisture.

Through the vents, the air is distributed throughout the space. Air filters might be installed on the vents to prevent dust and other tiny particles from entering the space. There are other locations within the ducting where air filters might be installed.

Heat Pumps vs. Air Conditioners

Using the same parts as air conditioners, heat pumps do the opposite procedure from air conditioning devices. As a result, during cold weather, heat from the exterior would be transmitted inside.

HEAT PUMP

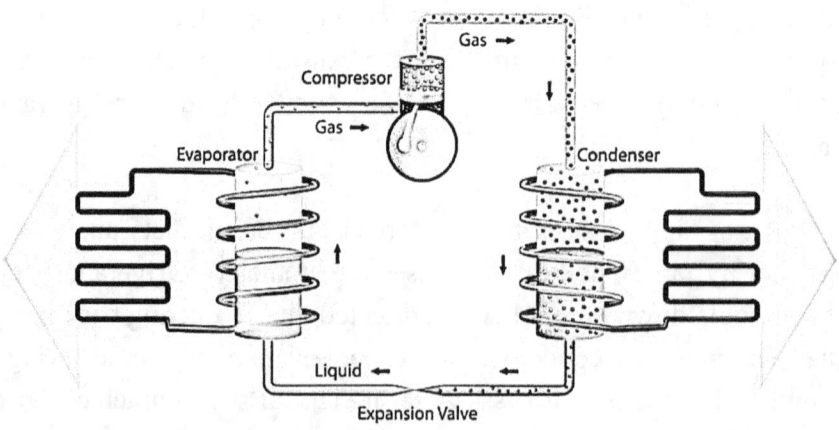

Split Units

To make clear how these HVAC system parts are utilized, the condenser/compressor is outdoors for split units, which are one of the most popular kinds of air conditioning/heat pump systems.

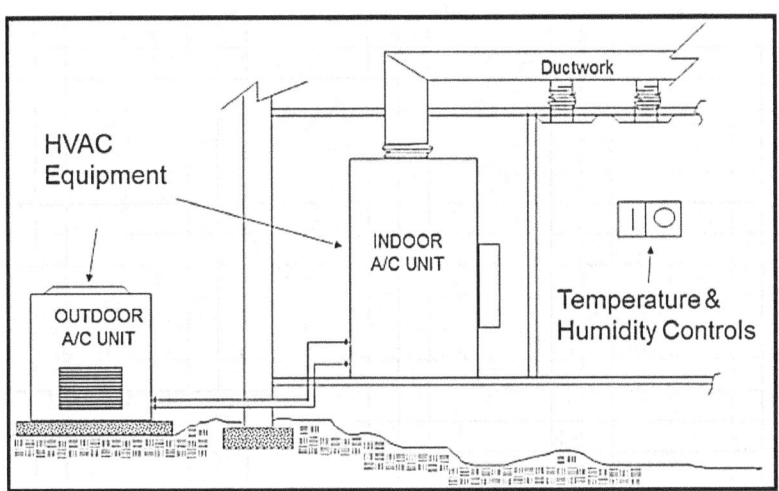

Working Principle of HVAC Systems

An outline of how HVAC systems function may be found below.

- The air return setup must be turned on for an HVAC system to operate. The air return system of an HVAC system acts as a point of entrance and exit for the ventilation cycle, which runs continuously while the system is in use. An exhaust fan is used to drive interior air outside whereas industrial axial fans are used to pull outside air into the space.
- The air filters filter out dust and other bacterial debris during the intake phase.
- Additionally, one of the next two tasks may be carried out by an HVAC system.

1. Air Heating

The HVAC system's heating unit has to be turned on to heat the air. Electronic heating components are used in the HVAC system to provide heat. Heating components include thermostats, induction coils, electronic heaters, and more. The heating element forms a heated zone in the route during suction air movement, and the air heats as it passes through it. Warm air is then introduced into space.

2. Air Cooling

The cooling unit is triggered to chill the air. Coils are used as a component of a heat exchanger in air conditioners to chill the air. There are two kinds of heat exchangers: shell-and-tube heat exchangers and cross-flow coils. Through the exchanger unit, which employs a refrigerant to remove heat from the suction air, only cooled air is fed into the room. A compressor included inside the cooling units liquefies the refrigerant.

- The space may be made colder or warmer by using a blower fan device.

Although heating or cooling the air itself is the method of humidification or dehumidification, spray humidifiers are often used in HVAC systems in areas with low humidity.

Chapter 3: Basic Engineering Principles Used in HVAC

Force [Newtons, N]

Force may be described as a push or a pull, put simply. It is the fundamental phenomenon of mechanical engineering and it applies to anything that can move a body, stop a body moving, or alter the direction of motion. Newtons [N], the unit of force, is named after Sir Isaac Newton, who in the seventeenth century was a pioneer in the study of force and motion.

Pressure [Pascals]

Pressure is defined as the force exerted per unit area.

One way to define it would be as the force intensity measured at any location on the contact surface. The pressure is constant at every place on the surface when a force is applied uniformly across the region. By dividing the total force applied to a surface by the total contact area, it may be computed.

Atmospheric Pressure [Pabs]

The atmosphere is an envelope of air that rises from the Earth's surface and envelops the planet. Because of gravity, air has mass and exerts a force known as weight. The pressure is the force per unit area. We refer to this force acting on the surface of the Earth as atmospheric pressure.

Gauge Pressure [Pgauge]

The majority of pressure measurement devices calculate the difference between a fluid's pressure and atmospheric pressure. Gauge pressure is the name for this.

Absolute Pressure [Pabs]

Total of gauge pressure and atmospheric pressure is absolute pressure.

Vacuum

When the pressure is less than the atmospheric pressure, the gauge pressure is said to be negative and the absolute pressure, or the point at which there is no air at all, is referred to as a vacuum.

The normal force per unit area applied to a real or imagined planar surface in a gas or fluid is known as pressure.

Density [ρ]

It may be expressed as the mass per unit volume or as the mass of a material divided by its volume.

Density(ρ) = mass(m) $\div$ volume(V).

The reciprocal of density, or volume per unit mass, is called specific volume (v).

v = V/m

The weight of a material divided by its volume, or the weight per unit volume, is known as specific weight (Ws).

Ws = m/V

Work

Work is considered done when a system moves as a result of a force acting upon it. The quantity of labor is equivalent to the force multiplied by the displacement component running parallel to the force. External work is defined as work done by or on a system when the system as a whole exerts a force on its surroundings and a displacement occurs.

Energy

Whelans able to do tasks, it is considered to have energy. More broadly speaking, energy is the body's ability to produce an impact.

Energy belongs to a class called

1. Consider stored energy. The potential energy stored in dams and chemical energy contained in the fuel

2. Transitional Energy, such as Heat and Work.

The three primary types of energy are internal, kinetic, and potential energy. Below is a description of the three types of energy.

When a body can do tasks, that body is considered to have energy.

Potential Energy

It is the energy that the system has stored as a result of where it is in the gravitational field. The energy needed to raise a heavy item, such as a construction stone, from the ground to the roof is stored in the stone as potential energy. For as long as the stone stays in its place, this potential energy is conserved.

$PE = mgH$

Where H = Height of object in Units Joules above the datum

Kinetic Energy

When a kilogram body moves at a speed of v m/s relative to the observer, the kinetic energy stored in the body may be calculated as follows: K.E. = 221 mv. As long as the body is moving at a consistent speed, this energy will stay stored inside it. Kinetic energy is also 0 when the velocity is zero.

Internal Energy

Mass is possessed by molecules. In both liquid and gaseous phases, they exhibit rotational and translational motion. These molecules have a lot of kinetic energy stored in them due to their mass and mobility. Since molecular velocity is a function of temperature, every temperature change also causes a change in molecule kinetic energy.

Additionally, some forces pull molecules toward one another. These forces are strongest when the molecules are in a solid form and tend to disappear when they are in a perfect gas state. It is vital to overcome these forces for a solid to melt or a liquid to evaporate. Potential energy is the energy that molecules store to effect this transition.

The whole energy of the body, whether it be chemical, nuclear, thermal, gravitational, or another form, is known as internal energy. The symbol "μ" represents the energy that is stored inside the body. The explanation given above makes it clear that measuring the internal energy's absolute value is impossible. On the other hand, we can quantify the internal energy changes. It is crucial to understand what causes the internal energy to change since thermodynamics deals with changes in the system's internal energy. The system's work, heat absorption or release, the addition or removal of materials, or any combination of these factors may all result in a change in internal energy.

Heat

Among the several types of energy is heat. The fact that heat can be transformed into other types of energy and that other types of energy can also be transformed into heat serves as evidence for this. It is widely acknowledged that heat is a molecular energy, and thermodynamics states that heat is the internal energy of matter.

Heat is regarded as the lowest kind of energy as it may be transformed into all other forms of energy. The temperature difference determines the amount of heat energy that may be used for work.

Heat Capacity

It may be described as the amount of energy required to add or subtract from one kilogram of a material to alter its temperature by one degree Celsius. Heat capacity is a phrase used in refrigeration technology to describe how much heat has to be evacuated to chill different types of items.

Sensible heat (QS)

Sensible heat is defined as heat that causes a temperature to change without altering its phase. When temperature changes, the equation determines the change in sensible heat.

$m \times CS\ (T2 - T1) = QS$ Note: The heat capacity at constant pressure is represented by CS. m is the substance's mass in kilograms; $(T2 - T1)$ is the temperature difference in degrees Celsius;

Latent Heat (Ql)

The heat at which a material undergoes a phase transition without experiencing a temperature change is known as latent heat. It is the quantity of heat needed to transform a substance's condition.

QL is equal to $mHCw(w2 - w1)$ Note that: m = mass of the material in kg; Cw represents the moisture's heat capacity; and $(w2 - w1)$ = change in moisture content in grams per kilogram.

Total Heat (Qt)

Sensible heat and latent heat add up to total heat. Measurements of heat are made above a certain datum. Water is solid below a certain temperature, hence these readings with it are at 0 degrees Celsius. As an illustration: The steam's sensible heat, latent heat, and total heat are shown below.

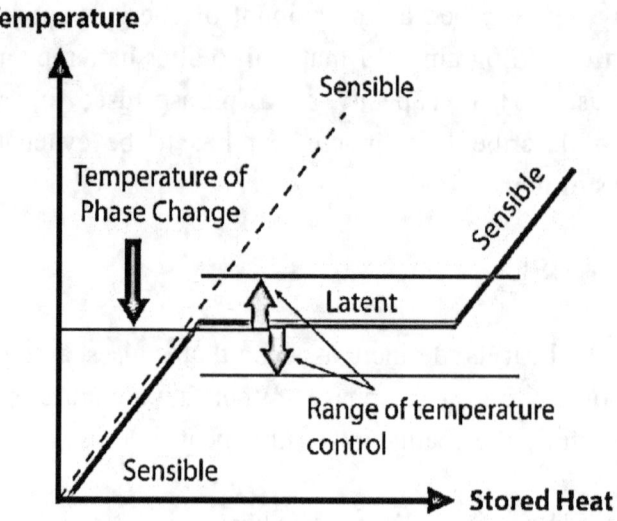

Temperature and its measurement

One of matter's properties is temperature. It is a measurement of the relative value and intensity of heat that is present in stuff. When a substance's temperature is contrasted with another reference temperature, it is referred to as hot or cold. A body is considered to be hot when its temperature is high, which is indicative of a high amount of thermal pressure or heat intensity.

Because heat has both amount and intensity, it may be measured like other types of energy. Although heat cannot be seen, it may be felt by different substances in different ways, either by altering their condition or by causing varying degrees of feeling when in touch with the human body.

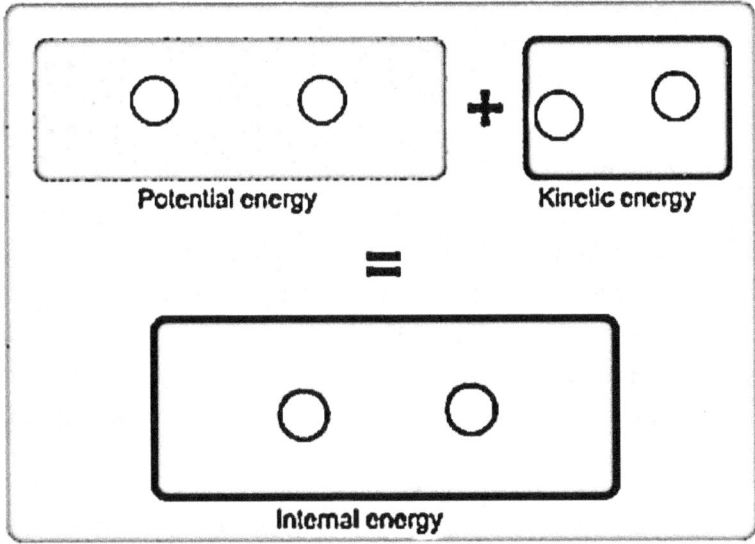

Since temperature is a measurement of heat content, it may be obtained by observing how heat affects various characteristics of matter, as follows:

• Heat addition causes a substance's volume or pressure to rise while maintaining a constant volume. Using a mercury thermometer, this feature is utilized to measure temperature.

• Metals' resistivity rises with temperature, and resistance thermometers make use of this property.

• A current flows in the circuit if two junctions composed of two dissimilar metals are kept at different temperatures. This characteristic is used in thermocouple measurement.

Substances undergo hue changes in response to temperature variations. Radiation pyrometers employ this characteristic to measure temperature.

Pressure and temperature relationship

At sea level, water boils at 1000 degrees Celsius when air pressure is applied to it. The boiling point rises when pressure is raised above air pressure, as in a deep mine shaft, and falls when pressure is lowered below atmospheric pressure, as in a mountaintop environment. Since water boils at a relatively low temperature when there is a vacuum, boiling water does not always need a high temperature. This also holds for other liquids, including different types of refrigerants. Except for having lower boiling point ranges, these refrigerants are identical to water in every way. Most refrigeration and air conditioning systems employ this pressure-temperature connection.

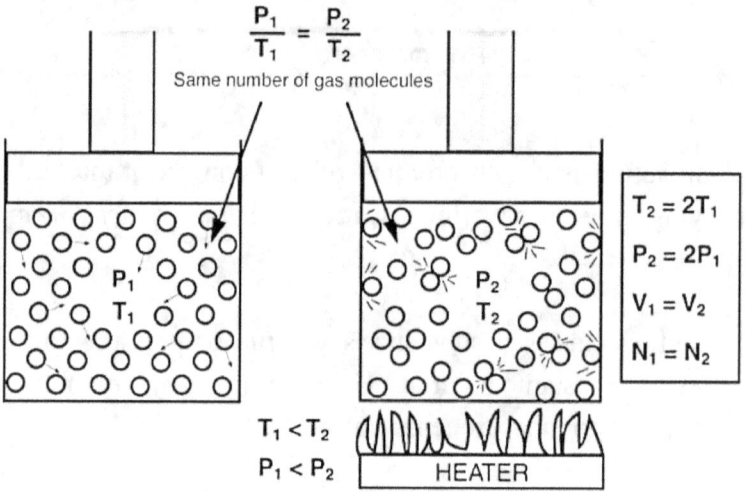

Chapter 4: Interpretation of Ratings and Adherence to Standards

Understanding the subtleties of ratings and adherence to standards is like reading the comfort blueprints in the complex realm of Heating, Ventilation, and Air Conditioning (HVAC), where the dance of temperature management meets the art of environmental mastery. We take a tour through the maze of industry standards in this chapter,

"Interpretation of Ratings and Adherence to Standards," which demystifies the language of ratings in plain English.

Picture yourself in front of a variety of HVAC systems, each with a label that seems to be able to communicate a secret code. These acronyms—SEER, EER, and AFUE—may seem confusing at first, but they are the key to maximizing the effectiveness and performance of your HVAC investment.

SEER: Decoding the Efficiency Quotient

First, let's talk about the Seasonal Energy Efficiency Ratio, or SEER. To put it another way, SEER calculates an air conditioner or heat pump's cooling effectiveness. The more effectively a system converts energy into cooling power, the higher its SEER rating. Think of it like an automobile's miles per gallon (MPG) rating: a higher SEER number indicates that your HVAC system is similar to a fuel-efficient car, providing more cooling for each energy unit used.

EER: A Snapshot of Energy Efficiency

The Energy Efficiency Ratio, or EER, provides a quicker perspective on efficiency. It's similar to capturing a momentary picture of the operation of your HVAC system. EER gives you information about how well your system performs during periods of high demand, while SEER looks at performance throughout an entire season. Consider it your system's efficiency measure on those sweltering summer days when it is working nonstop.

AFUE: Unraveling Heating Efficiency

AFUE, or Annual Fuel Utilization Efficiency, becomes our road map when we turn our attention to heating. This grade indicates how well your boiler or furnace converts fuel into heat throughout the heating season. Consider AFUE to be the portion of fuel converted to useful heat.

When the AFUE rating is 90%, 90% of the fuel is converted to heat and the remaining 10% is released as exhaust.

Understanding ENERGY STAR® Certification

Let's now discuss the industry gold standard, which is ENERGY STAR® certification. This designation indicates that a product—whether it be a heat pump, furnace, or air conditioner—meets the stringent energy efficiency standards established by the Environmental Protection Agency (EPA) of the United States. It's as if your HVAC system's exceptional energy conservation and greenhouse gas emission reduction are verified.

Adherence to Standards: Your Assurance of Quality

The foundation of HVAC dependability is standardization, not ratings. Seek for systems compliant with industry standards to make sure your investment is in line with industry standards set by groups such as the American National Standards Institute (ANSI) and the Air Conditioning, Heating, and Refrigeration Institute (AHRI). Because of our dedication to standards, you can be sure that your HVAC system is built to function as advertised on the label.

BOOK II: HVAC INSTALLATION MASTERY

Introduction

The crucial scene in the HVAC play is installation, which is when possibility becomes performance and the theoretical gives way to the real. This part delves into the complexities of HVAC Installation Mastery, taking you through the many HVAC system landscapes, the painstaking installation procedure, and the skill of achieving maximum efficiency right from the start.

Imagine the canvas of your living area changing as we set out on this leg of our HVAC adventure, with every aspect finding its place like a practiced group prepared to perform its role. Our trip promises to show the route to a properly tuned and harmonious HVAC system, from the grandeur of extensive installations to the precision needed to enhance efficiency.

Chapter 1: Types of HVAC Systems

Central HVAC Systems

An example of a building automation system that regulates the temperature and humidity across a complex or building is a central HVAC (heating, ventilation, and air conditioning) system. The cost of installation for central systems is often higher than that of standalone HVAC units. Nevertheless, since they control the environment in many places, they may eventually save money on energy costs.

Installing a central HVAC system across your building has several advantages over employing separate units:

1. Central systems lessen the possibility of hot and cold patches by regulating temperatures more uniformly across a facility. This guarantees that everyone in the facility has a pleasant atmosphere

and keeps the equipment from being overworked and prone to malfunctions.

2. Central systems will not accidentally overheat or undercool regions since they are usually more accurate than standalone units. Furthermore, patterns in space (such as dense populations of people or furniture) that would be difficult for individual unit sensors to identify are often picked up on by centralized systems.

3. Centralized systems can diagnose issues fast, even during peak hours, and adjust their routing in response as they manage so many various components of the heating, ventilation, and air conditioning grid at once. This enables them to reduce downtime and maintain seamless company operations regardless of external events!

Portable HVAC Systems

Those who want to heat their houses but lack the space or funds for an underground system might choose portable HVAC systems. Portable units are more energy-efficient than conventional in-ground systems since they can be moved from one area to another. Not only are they simpler to install since they don't need excavation or electrical work, but they are also often less expensive than standard in-ground units.

There are now several solutions available if you're searching for a portable heater that will satisfy your unique heating requirements. Oil furnaces, solar-paneled air conditioners, and gas heating pads are a few of the most well-liked portable HVAC systems. For a flawless installation, make sure you thoroughly read the manufacturer's instructions before proceeding with whatever choice you make!

Heat Pumps

A heat pump is a kind of air conditioner that transfers heat from one place to another using the laws of thermodynamics. This procedure transfers thermal energy from the source, like an interior furnace or solar panel, to the destination, such as outdoor air, by use of a refrigerant, which is a material that absorbs and stores heat.

A heat pump may be an excellent option in frigid areas, where it's often too warm for window air conditioners but too cold for central heating. They are ideal for workplaces and big residences since they consume less power than conventional AC units and cool huge areas fast.

The biggest drawback of heat pumps is that they don't always work well to heat tiny areas; on average, they use three times as much energy to provide the same amount of warmth as a window air conditioner of the same size. However, a heat pump can be the ideal solution if you're searching for a cost-effective approach to chill your house or place of business in a cold area without consuming a lot of power!

What makes purchasing an electric heat pump worthwhile?

Air Conditioning Units

Air conditioners often have a difficult time operating during the heat. People prefer to spend more time inside due to the high temperatures. This implies that to keep everyone cool and comfortable, air conditioners have to work more than normal. But there are a few things you can do this summer to put even more strain on your air conditioner.

- When the weather is hot, keep the windows closed. This lets the air conditioning from the unit in and keeps the rooms cooler.

- Apply evaporative cooling: This technique cools surfaces by drawing moisture from the air, as opposed to utilizing electricity or flowing water. If you don't have access to an air conditioner or it's too hot outdoors for it to function effectively, this is useful. Simply lay damp towels against window panes and let them draw moisture from the surrounding air. This will result in liquid droplets that when in touch with variations in ambient temperature, disperse heat.

- Set your thermostat as high as you can: Increasing the temperature by 10 degrees Fahrenheit (5 degrees Celsius) may result in a 30% reduction in energy use. Recall that in warm weather, turning up your thermostat too much could make you uncomfortable.

Fan Coil Units

An air conditioning system that employs fans to distribute air about the space is called a fan coil unit (FCU). Because they produce no pollutants and make very little noise, they are often seen as being more ecologically friendly and efficient than traditional AC units.

What advantages can FCUs offer?

- They often use less energy than conventional air conditioning systems.

- They may lower your electricity costs since they use less energy.

- Fan coils are great for bedrooms and other places where you desire quiet since they are often quieter than conventional AC units.

Mini Split Units

The many advantages of mini-split units are contributing to their growing popularity. With the help of these small systems, you can keep your house comfortable while lowering your carbon footprint by heating and cooling it with renewable energy.

Some of the most popular advantages of mini-split units are as follows:

It's simple to install them. You may immediately begin heating or cooling your room by connecting it to an electrical outlet!

They're reasonably priced. Building a mini-split is often far less expensive than building a full-blown air conditioning or heating system, with most costing between $1,000 and $5,000.

- They use less energy. This implies that micro splits function effectively even with low electrical sources since they don't need a lot of it to run.

- They aid in lowering air pollution. Large volumes of pollutants are produced by heating and air conditioning systems, including dangerous particles known as ozone-depleting substances (ODS). Without compromising comfort or convenience, you may significantly cut these emissions by switching to micro splits from regular units!

High-efficiency Furnaces

Today's market is filled with a variety of high-efficiency furnace models. Which is best for your company?

Low-NOx Furnaces: One of the main causes of air pollution, nitrogen oxide (NOx) is produced at reduced levels in these furnaces. They are a great option for small companies and households without prior furnace installation or maintenance knowledge since they are less costly and simpler to use than other kinds of furnaces.

High-Efficiency Furnaces: Compared to low-NOx furnaces, these furnaces generate more heat and power, but they are also often more costly. They may save energy expenses by up to 30%, making them ideal for big commercial buildings or industries where efficiency is crucial.

Dual Fuel Furnace Systems: This kind of furnace combines high-efficiency and low-NOx technology to obtain the best possible performance level based on the particular requirements at hand. Although dual fuel systems often cost more than either low-NOx or high-efficiency systems, they provide additional alternatives for heating.

Chapter 2: The Comprehensive Installation Process

A new home or building's HVAC (heating, ventilation, and air conditioning) system installation is a challenging procedure that involves numerous considerations. It can also be necessary, depending on local laws, for HVAC installations to be performed by certified, highly trained HVAC specialists.

Nevertheless, for a thorough, step-by-step guide to HVAC installation, continue reading if you're curious about the procedure, whether you want to perform some maintenance on your present system, or would just like to know more about what to anticipate during the installation process.

HVAC Installation for New Construction

The process of installing HVAC systems in a new building comprises several intricate steps that, when completed by qualified experts, can usually be completed in a reasonable amount of time.

The whole operation may be finished in three to five days, depending on the size of the new structure, the number of personnel on the HVAC installation team, and climatic concerns.

Planning

An HVAC specialist must visit the location to personally inspect the building site's layout and collect precise measurements to arrange the ductwork's placement before starting the HVAC installation.

One of the most crucial phases in the HVAC installation process is planning, as this will define the overall effectiveness of your system and have a significant impact on the entire cost of doing the work.

Your HVAC expert must carefully consider the supply and return duct routes when designing the ductwork throughout your home to minimize the distance that your hot and cold air must travel. The system's efficiency decreases with the distance the air must travel. Additionally, it is best to minimize curves and twists to lessen airflow impediments. This will maximize temperature management while allowing your warm and cold air to circulate freely around your home.

Organizing your ducting well will also help to minimize temperature differences between rooms. Though the rest of the home was certainly pleasant, you have undoubtedly lived in a place where there was a room or region that was never hot or cold enough. Numerous variables, like the location of the thermostat and the effectiveness of the airflow, may contribute to this.

To guarantee that the HVAC thermostat receives a reading that appropriately represents the temperature throughout the house, it must be positioned in the center of the house. The room closest to your heater, for instance, will be among the first to get warm air from the supply duct if you place the thermostat there. The room will heat up faster than other parts of the house as a result, and the thermostat will shut off the HVAC system before it can sufficiently heat or cool other parts of the house when it detects that the room has achieved the desired temperature.

The installation of dampers in key locations around the house is one method to help with this problem. Dampers are installed inside the ductwork of your newly constructed home to control the airflow throughout the house. Although there are many varieties of dampers, "balancing dampers" are the most often utilized kind in residential settings.

To regulate the amount of air that travels to specific locations, these dampers are placed at branch points inside larger ducts. Your HVAC specialist will install a balancing damper at the branch of your ductwork that distributes air movement to a smaller area, for instance, to lessen the quantity of airflow to this room and avoid overheating or cooling it down.

Additionally, some houses feature HVAC systems that are known as "zoned." A zonal HVAC system consists of a central control panel that oversees the whole system in addition to discrete thermostats placed at separate places. This makes it possible for the HVAC system to keep various parts of the home or building at different temperatures. Generally speaking, zone damper systems are motorized to adapt to the various thermostat settings.

The placement of supply and return points is a crucial factor to take into account when designing an HVAC system's ductwork. Well-versed HVAC specialists will locate supply and return points by their knowledge of a home's natural air circulation patterns.

If you are not acquainted with these terms, the return is the place where the air is brought back into the HVAC system from the conditioned regions, and the supply is the place where hot or cold air flows out to adjust the temperature inside the room or area.

To guarantee that the HVAC system is operating well, the return system is essential. The reason supply and return points are often located on opposing sides of a room is that air drawn into the return duct provides a negative pressure system that aids in drawing the conditioned air into the space.

Air filtration is one of the HVAC return system's key functions. A filter is most likely within the return duct if you have ever removed the grille, which is the metal covering with large, horizontal holes. If you were unable to locate a filter here, you need to install one as this will extend the functionality and life of your HVAC system.

Moisture, dust, hair, and other particles that can be flowing throughout your house are removed by HVAC return filters. You may lessen the accumulation of impurities inside your heater, air conditioning, and ductwork by making sure that clean, filtered air is sent back into your HVAC system. Maintaining the cleanliness of these systems' internal parts will prolong their lifespan, enhance their functionality, and need less maintenance. Additionally, this will aid in raising the standard of air being circulated in your house.

Depending on the kind of filter you're using, return filters may be updated every 30 to 90 days and are reasonably priced. Other factors, including the quality of the filter and the location in which you reside, will also have an impact on how long your filter lasts. For an air filter, it's a good idea to replace it if it seems to be unclean.

Your HVAC professional will design the ductwork arrangement for your HVAC system while also taking into account other elements including the size of your home, the local temperature, the R-value of your insulation, and any applicable zoning rules or restrictions.

These are all significant attributes that will establish your total HVAC installation requirements.

Ductwork Installation

Your technician will start installing the ductwork throughout your newly constructed home after finishing the design stage of your HVAC installation. To provide the required framework for securely installing your ductwork, this step of the procedure must be finished after the whole home is constructed. This will often happen after the installation of the electrical system and wiring in the home.

Installing ductwork after the electrical components is the best way to minimize conflicts in the layouts of the two systems since ductwork is huge and may take up a lot of space between the joists and studs on your floors and walls. Also, the presence of electricity will enable HVAC specialists to test and turn on the system and ensure that its final settings work.

Depending on the precise function and location of the ductwork, HVAC experts will employ a range of duct diameters and shapes during installation. You have most likely encountered a mix of round and flat ducts if you have ever been inside an unfinished home or basement. Flat, round, or oval-shaped ductwork are also available. Flexible ducting may be twisted and molded to fit in spaces that are difficult to access.

Though supply ducts are usually considerably smaller than return ducts, all of these solutions are appropriate for both types of ducts.

Additionally, ductwork may be constructed from a wide range of materials, including copper, stainless steel, aluminum, and galvanized steel. However, galvanized steel and aluminum are the most often used materials in dwellings.

Because of its exceptional durability and zinc coating that prevents corrosion, galvanized steel is a popular material for residential ducting. Because of its superior corrosion resistance, aluminum may also be utilized in areas where humidity is a major problem. However, it is not uncommon to see a home's HVAC system combine the two materials.

Although they may be used in HVAC installations, copper and stainless steel are less popular because of their greater cost. Stainless steel and copper are usually exclusively used in specialized applications that ask for particular qualities.

Insulation is often used to cover ductwork to stop undesired heat transfers and to stop moisture from building up outside the ducts. Because it may harm other parts of the house, condensation can be a serious issue in certain places. For instance, water stains might appear on drywall if condensation is allowed to build up on a duct behind a section of the material.

Everywhere two sections of duct intersect, ductwork has to be sealed off with duct tape in addition to being insulated. By doing this, an airtight fit will be produced, increasing the system's total airflow efficiency.

System Installation

Your HVAC specialist will install your heating and cooling equipment after finishing the duct system. Heaters in residential HVAC systems are always installed inside, while air conditioners consist of two independent parts, one of which is located outside.

The evaporator coil will be on top of the heater so it can feed cold air to the home's supply pipes, and the condenser component and exhaust fan will be outdoors, often situated directly next to the house.

Your HVAC specialist will have already installed the required gas lines and ducts to link your heating and cooling units, and they will have also decided on each unit's location.

Heaters

Typically, home heaters are kept in a utility room, however, they may also be kept outside in an unfinished basement.

It is imperative that you cut off the gas and the power before starting any heating installation. Even though they are often already turned off for the whole home while the building is underway, you should still double-check when installing HVAC systems in new construction. To prevent any problems arising from competing projects, inform any other teams who are working in the house while you install the HVAC system.

If it is constructed according to code, heaters will always have their circuit. Even though circuit breakers in new construction are often labeled clearly in the breaker box, it's a good idea to double-check that you have the appropriate breaker just in case anything was labeled incorrectly. To test the heater switch, take the switch plate off of the on/off switch on your heater and check with a voltmeter to determine whether the switch is getting electricity. Although multimeters are simple and affordable instruments, if you are not acquainted with their operation, spend some time watching a video to learn the correct way to use them. It is recommended to cut off the power to the whole home if you are unable to test the switch.

There ought to be a gas line valve close to the location of the heater installation. Before starting work, make sure the valve is in the off position. When the valve is in the open position, its handle will be parallel to the gas line; when it is in the off position, it will be perpendicular to the line.

You may start attaching the heater to your supply and return ducts once you have confirmed that all of your gas and electrical sources are off. To finish this phase of the HVAC installation procedure, you will need some extra supplies and equipment. This is because this part does call for some customization.

First, you must make a hole in the heater's side to attach your return line to it. Since return duct sizes vary from home to home, heaters are not produced with precut openings. Before making any cuts into the sheet metal on the exterior of your heater, be sure you are attaching the return duct to the right side by consulting the installation handbook for your heater.

A blower motor located within the heater's base draws air in and pushes it upward into the supply duct. To fit the ductwork of the house where the heater is being put, heaters come with blowers that face both left and right. Lifting the heater's front cover is another approach to make sure the blower is pointing in the right direction. You will need to replace your blower with a heater that is oriented correctly if it is not facing the same side as your return duct.

Once the return duct's proper position has been determined, cut a hole in the heater's side that is the same size and form as the return duct. It's usually rather simple to attach your return duct and doesn't need any specific equipment. Just butt the return duct's edges up to the opening, then use duct tape to seal the joint. Additionally, there ought to be a hole in your return duct where you may occasionally replace the air filter. Before you turn on the heater, make sure you place a filter in this area that is big enough to cover the aperture entirely.

The gas pipes to your heater must then be connected. The threads on the gas line connectors of your heater aren't usually airtight, so you'll need to seal them with a specific sealant. After sealing the threads on your gas line, connect the flexible gas hose from the gas line valve to the gas valve on your heater.

The electrical wires from your heater must be connected to your switch after your gas line has been firmly attached to the gas valve. Two wires from your new heater will travel to the location where your switch will be mounted. Use a wire nut or another kind of electrical connector to join the neutral (white) wires in a matching manner. Additionally, confirm that the green ground wire is attached to the ground screw located within the switch box. Following the connection of your neutral and ground wires, wrap the black live wire (from your heater) over the switch's side screw to secure it. Repeat with the black wire from your switch. Your heater will be able to turn on when the switch is flicked because the circuit between the two black wires will be complete.

The thermostat wires must be attached to the heater's circuit board once the main power supply is turned on. You must consult the wiring diagram in your installation instructions before proceeding with this step. Make sure that all of the wires, including those for the thermostat and the AC condensing unit, are connected to the appropriate terminals on the circuit board.

You must attach the ducting to the top of your heater after the electrical connection is complete. You may need to put off finishing this last step since the evaporator coil for your air conditioner will often be installed on top of the heater in most house HVAC systems.

Customization is necessary for attaching the ductwork, so measure well before cutting and arranging the walls to connect to your duct. While there are a few methods for doing this, making sure the fit is airtight is the most crucial aspect of this process. The bottom part is screwed straight into the top of the heater or evaporator coil box, while the higher half is usually connected with a section of S channel. Use duct tape to seal everything after your ducting is connected.

Once the installation of your heater is finished, you should check your system's gas and air pressure. It is recommended to consult your installation instructions to check the manufacturer's suggestions since this procedure may call for certain specialist gear. To ensure that your system is operating properly, it would be wise to leave this task to a professional.

Air Conditioning

You must take off the panel on the outside unit that is above the gas lines before installing your air conditioning system. There are two main components to the outside unit. The refrigerant is compressed and condensed by a condenser unit first. The second component helps dissipate the extra heat produced during the compression of the refrigerant: a fan.

The panel containing the gas valve's electrical connections is where the unit's electricity is located. The location of the electrical and ground wires should be indicated in the installation instructions provided by the manufacturer, but other than that, it's a simple operation that should be easy to finish. You must connect the cables for your thermostat after you have connected the wires for your electricity.

You will need to solder your gas lines after installing all of your wiring. One big gas line and one tiny gas line will be present. Your outside installation will be finished after you solder them to the matching gas valves on the outdoor unit. The valves must remain closed until the installation is finished in its entirety.

When purchasing a new air conditioner, the system is often pre-charged with refrigerant. As of right now, your gas lines are not sealed since you have only completed half of the installation. You must put the evaporator coil above your heater before you can open the valves.

Your evaporator coil has to be installed above the heating unit. Your supply ducts should have previously considered this in the event of new construction, giving you adequate space to put your evaporator coil above your heater.

Self-tapping screws are required to fasten the metal box that houses your evaporator to the top of your heater. The box is provided with your heater. Using a technique similar to the one we discussed when attaching your heater to the supply duct, you must attach the top of the box after it has been screwed down.

Duct tape and duct sealant should be used in conjunction to seal up the interior joints. Use a paintbrush to apply duct sealant and let it cure after using duct tape to close up the seams.

You may change the front panel and install the evaporator coil once your evaporator box is connected to the duct. You may solder the gas lines that connect your outdoor unit to the evaporator coil after replacing your front panel.

You may now go back to the outside unit and open the gas valves, letting the refrigerant circulate throughout the air conditioning system, now that everything is installed. After that, you may turn the system on, but to make sure it's operating properly, you might need to make a few more adjustments.

Chapter 3: Establishing Optimal Efficiency During Installation

Energy efficiency in HVAC (Heating, Ventilation, and Air Conditioning) systems has to be maximized for US homes and businesses. Energy-saving measures may be put into practice to cut down on energy use, save utility expenses, enhance interior comfort, and support environmental sustainability. The practical advice and techniques to maximize energy efficiency in HVAC systems will be discussed in this article. To achieve energy efficiency, every step counts, from using cutting-edge technology and doing routine maintenance to incorporating building inhabitants and properly sizing the facility. Businesses and households may see long-term financial savings, enhanced indoor air quality, and decreased environmental impact by putting these strategies into practice.

Understanding HVAC System Energy Efficiency:

It is crucial to comprehend how HVAC systems use energy as well as the applicable energy efficiency requirements and ratings to maximize energy efficiency in these systems. In the US, HVAC systems are mostly responsible for the energy used in home and commercial buildings. Businesses and families may save a lot of energy and help create a greener future by increasing the efficiency of these systems.

Important Techniques for Enhancing Energy Efficiency

1. Appropriate Sizing and Load Calculation of HVAC Systems:

It's crucial to size HVAC systems appropriately depending on the unique heating and cooling requirements of a structure. Reduced efficiency and energy loss might arise from either an oversized or undersized approach. Determining the right system size involves doing precise load calculations and taking into account elements like insulation, building orientation, and occupancy.

2. Consistent Upkeep and Cleaning:

Maintaining HVAC systems regularly guarantees peak performance and energy economy. Essential maintenance activities include examining and cleaning coils, calibrating thermostats, cleaning or changing air filters, and monitoring refrigerant levels at regular intervals.

3. Investing in Energy-Efficient Technology:

Think about replacing outdated, energy-inefficient HVAC systems with more recent, energy-efficient alternatives. Advanced features including variable speed motors, sophisticated controls, and better insulation are often seen in energy-efficient HVAC systems. Over time, these changes may save a substantial amount of energy.

4. Putting Smart Thermostat Controls Into Practice:

With smart thermostats, you can schedule, remotely monitor, and precisely adjust the temperature. By adjusting temperature settings according to occupancy and activity patterns, smart thermostats can save energy and increase comfort.

5. Improving Ventilation and Air Distribution:

Effective ventilation and air dispersion are essential to the effectiveness of HVAC systems. Utilizing demand-based ventilation systems, ensuring balanced airflows, and caulking ductwork to stop leaks may all assist maximize energy efficiency and indoor air quality.

Energy-Saving Practices for HVAC System Operation

Using energy-saving techniques in the regular operation of HVAC systems, in addition to the crucial tactics already stated, may further maximize energy efficiency:

1. Optimization of Temperature and Setback:

Considerable energy savings may be achieved by adjusting temperature settings according to occupancy and putting setback techniques into place during off-peak hours.

2. Making Use of Daylighting and Natural Ventilation:

Reducing the need for artificial lighting and cooling by making use of daylighting methods and natural ventilation may save energy.

3. Effectively Controlling Humidity Levels:

Maintaining a building's humidity levels is crucial for both energy efficiency and tenant comfort. Controlling humidity effectively lessens the strain on HVAC systems and enhances their general functionality.

4. Making Certain Correct Sealing and Insulation:

Buildings with adequate insulation and sealing limit heat transmission, cut down on energy loss and boost HVAC system performance.

5. Air Filter Monitoring and Adjustment:

It is possible to increase energy efficiency, lessen system stress, and improve airflow by routinely checking, cleaning, or replacing air filters.

Advanced Technologies for Energy Efficiency

Technological developments have brought out several options to further maximize energy efficiency in HVAC systems, including:

1. Modulating controls and drives with variable speeds:

Variable speed drives provide precise control and increased energy efficiency by varying the speed of HVAC system components in response to demand. Gradual modifications are made possible by modulating controls, which guarantee peak performance.

2. Systems for Energy Recycling and Heat Recovery:

By putting energy recycling and heat recovery systems in place, waste heat may be captured and used, which reduces energy consumption and boosts system performance.

3. Energy management and building automation systems:

Building automation systems optimize energy use based on real-time data and preset settings by integrating multiple HVAC, lighting, and other systems.

4. Occupancy and Ventilation Sensors Based on Demand:

Demand-based ventilation systems increase energy efficiency by modifying airflow in response to occupancy levels. By turning on or off HVAC systems in response to variations in occupancy, occupancy sensors aid in the optimization of energy use.

Greetings, Reader

I hope you are doing well as I write this. We appreciate you taking the time to read The Updated HVAC for Beginners 2024. I value your opinions much and would be interested in knowing what you think of the book. Your viewpoint affects whether you have begun, are halfway through, or have done reading.

I really value your input, which will help me improve my next efforts.

We appreciate you taking the time to comment on The Updated HVAC for Beginners 2024. Your support means the world to me.

Enjoy your reading!

BOOK III: HVAC MAINTENANCE AND TROUBLESHOOTING

Introduction

Maintenance is the unsung hero that keeps the smooth rhythm of heating and cooling going strong. It is the quiet protector of our HVAC systems. We delve into the crucial procedures that maintain the effectiveness and dependability of your system as we examine the nuances of HVAC maintenance and troubleshooting in this chapter. Our path promises to provide you with the skills to become the conductor of your comfort symphony, starting with daily routines and ending with the empowerment of do-it-yourself repairs.

When we turn the pages of this chapter, picture the scene: your HVAC system performing at its best, skillfully adapting to the signals of the changing seasons. We cordially welcome you to join us on this journey through maintenance and troubleshooting, which will reveal the preventative steps, diagnostic insights, and practical solutions necessary to make sure your HVAC system continues to be more than just a fixture but a dependable partner in the comfort management of your house. Here, at the center of HVAC care, maintenance, and troubleshooting reveal the hidden details that provide continuous peace of mind.

Chapter 1: Regular Maintenance Practices

1. Arrange for expert preventive maintenance for your HVAC system.

A seasonal HVAC tune-up should be scheduled twice a year: once for the heating system in the autumn and once for the air conditioning in the spring. HVAC installation providers and professionals will completely service, examine, and troubleshoot the system during periodic maintenance checks to keep it operating properly and avoid failures. The HVAC specialist will

- Verify the settings and calibration of the thermostat.
- When necessary, tighten electrical connections.

- Make sure all moving components are lubricated.
- Examine and clean the condensate drain as necessary.
- Examine the controls inside the system.
- Tighten and clean the blower's parts.
- Clear the coils on the condenser and evaporator.
- Verify the charge of refrigerant.
- Inspect the connectors of the fuel line.
- Examine the heat exchanger, burner combustion, and gas pressure.

2. Change the Filters

To stop these contaminants from spreading throughout the house, filters collect and remove hair, dust, and other particles from the air. Replacing the HVAC filter once every thirty days will help you save energy and enhance the quality of air inside your house. Reliable heating undercooling is made possible by clean filters, which enable more air to flow through. Maximum airflow and filtration efficiency are best balanced in most systems by using filters with MERV ratings between seven and 13.

3. Evaluate the HVAC system visually.

Every month, check the system to identify any possible issues when you change the filter. Examine the registers and returns, the thermostat, the unit's exterior and inside. You should also look into the following in addition to that:

- Verify the thermostat's battery condition.
- Verify that the condensate system is draining correctly by looking at it and making sure the cabinet door and filter access are safely closed.
- Verify that the flue system is completely intact and firmly connected.
- Verify that all returns and registers are open and unblocked.

- Examine each register for evidence of mold.
- Velevellevelness of the outside unit. If needed, use rot-proof shims to level it.

4. Clear the Area Near the Indoor HVAC Unit of Clutter

Air quality and safety are enhanced when the space around your indoor HVAC unit is kept free. There is more surface area to gather dust that will ultimately find its way into the vent system the more items laying about. Additionally, clutter lowers local air circulation, which is detrimental to system performance. Additionally, clutter may make it more difficult to undertake maintenance and repairs and turn into a trip and fire danger.

5. Maintain a clear and clean outside unit

It is easy for fallen leaves, twigs, grass clippings, and other debris to gather near the outside HVAC unit. Every time you do yard care, clear the area surrounding the unit of any debris and give it a thorough hose-down if any dirt starts to build up. To assist ensure enough airflow, keep surrounding plants cut back at least two feet from all sides of the unit.

6. Control the Temperature in Your Home

Maintaining your HVAC system involves setting it to comfortable settings and minimizing its use when you are asleep or away from home. If you want the temperature to change automatically throughout the day, think about installing a programmable thermostat. The system will operate less often, use less energy, and last longer if you let the house remain warmer in the summer and colder in the winter while you're not home.

7. Replace the Thermostat's BaSome thermostats that are hardwired into the electrical system of homes. Others run on batteries. To avoid issues, replace batteries at least once a year.

8. Keep the carbon monoxide detector serviced

For residences using fuel oil or natural gas for combustion-based heating, a carbon monoxide detector is a necessary safety equipment. The alert might save your life in the case of an exhaust leak, poor ventilation, excessive gas flow, problem.

The typical lifespan of these gadgets is seven years. Every month, check whether the carbon monoxide detector is functioning properly by testing it; replace it if needed. Schedule a battery repment for every six months.

9. Keep an eye on your energy bills

A problem with your HVAC system may be indicated if you see a sudden surge in energy consumption or a steady increase while use remains constant. To get the system checked, arrange a service appointment with your reliable heating and cooling firm. Possible causes include low refrigerant, duct leaks, dirty filters, broken components, and other problems.

10. Take into Account Complete Replacement

An HVAC system may last for 15 to 25 years on average. That timetable may be affected by several factors, such as the kind of system, its brand, and how often it is maintained. You can extend the life of your heating and cooling system and keep it operating at peak performance for longer by giving it regular maintenance.

Chapter 2: Identifying and Resolving Common Problems

Issue 1: Insufficient cooling or heating

- To make sure the thermostat is set appropriately for the intended temperature, check the settings. Adjust the settings as needed to ensure your comfort level.

- Check the air filters and clean them if you find any that are dirty or blocked. There are situations when a full filter replacement could be more advantageous. The system's ability to cool or heat is increased by improved airflow from clean filters.

- To guarantee ideal ventilation, clear the space around vents and registers of any obstructions. Verify that nothing is obstructing the airflow, such as curtains, furniture, or other objects, since this might interfere with the system's ability to sufficiently heat or cool the area.

- Upgrade to a programmable thermostat for precise temperature control and increased energy efficiency. These thermostats allow you to configure several temperature schedules for different times of the day, ensuring optimal comfort with little energy use.

Issue 2: Weird noises

- Identifying the kind of noise and its possible source is the first step to take if your HVAC system is producing strange sounds. Is there a rattling, grinding, or screaming sound? Determining the source of the noise might help pinpoint the issue.

- Look for any loose fan blades, belts, or motor mounts in the HVAC system. If you locate any, gently tighten them to eliminate the cause of the noise. Unusual sounds and vibrations may be produced by loose components when they are operating.

- Lubricate your HVAC system's moving parts using the proper lubricants. This is relevant to components that need lube in order to operate correctly, such as fan motors and bearings. Proper lubrication prevents friction and noise from being produced by parts rubbing against one another.

- If the strange noise persists after doing the previously suggested steps, it may be best to see a licensed HVAC technician. They possess the expertise and know-how to recognize and address complex issues that can need more involved fixes or debugging.

Issue 3: Poor airflow

- Examine and clean every air vent and register in your home or building first. Remove any obstructions (furniture, dust, items) that might be preventing the airflow. Verify that all of the registers and vents are fully open and free of obstacles.

- Inspect and clean the condenser and evaporator coils of the HVAC system. Over time, these coils may accumulate dust and debris, preventing the best possible ventilation. To remove any buildup, carefully clean the coils with a soft brush or a vacuum equipped with a brush attachment.

- Verify that the blower fan is functioning properly. If the fan is not operating at its peak efficiency, try adjusting the speed. If you want to make sure the blower fan is operating correctly, see the manufacturer's instructions or consider hiring an expert.

- If the airflow is consistently inadequate, consider installing a duct booster fan to help. Duct booster fans are designed to increase airflow inside a specific ducting system, improving comfort and ventilation in general.

Extra Resources and Troubleshooting Advice

In addition to the thorough troubleshooting steps described for common HVAC problems, consider the following additional points and ideas:

- Check to make sure the power supply for your HVAC system is safely connected and working.

- Look for and repair any refrigerant leaks since low refrigerant levels may have an impact on system performance.

- Examine and clean the outside unit, making sure to get rid of any dirt, leaves, or grass clippings that can obstruct airflow.

- Verify that the fuses and circuit breakers attached to your HVAC system are working.

Chapter 3: Step-by-Step DIY Repairs

When it comes to doing your own DIY air conditioner repair at home, you need to be very careful, have the correct components and equipment, and know how to diagnose service faults.

To fix your air conditioner, get in touch with a local, professional HVAC expert if any of these items are missing.

Always keep in mind that you'll need tools and that you should feel at ease utilizing them.

If you're not comfortable doing repairs on your own, you should consider contacting a local, professional HVAC contractor. Repairing air conditioning equipment may be risky.

Sometimes it's required since the repair needs specialized equipment, which can be costly for a homeowner to purchase.

You'll need the appropriate equipment if you want to do more do-it-yourself projects around the house.

Nothing completes a task more quickly for a professional contractor than the use of power equipment. So which are the appropriate ones?

A Few Tools Will Be Needed to Get Started.

- 1/4″ nut driver
- Voltage tester or CAT III Multimeter
- Insulated screwdriver set
- Socket setand
- Pliersnosesedle nose will work the best
- Cordless drill
- Adjustable wrench
- Additionally, you may need slip-joint pliers, sometimes known as channel locks.

This is a list of the necessary components and materials.

1. Capacitor (to purchase this, you'll need to know the model and serial number of your air conditioner)
2. A compressed air or water hose condenser fan motor (to order this, you'll need to know your unit's model and serialThe contractor
3. Contactor (to purchase this, you'll need to know the model and serial number of your unit)

4. Fuses (or, perhaps, a circuit breaker)

How to Fix Your Home's Central Air Conditioner on Your Own

Issues with Services and Their Fixes.

When your air conditioner brightens in the middle of July, when it's the hottest time of day, what are your options?

You may work with an air conditioning system repair firm or a professional HVAC contractor.

Alternatively, if you think you're handy and have the necessary equipment, you may own an air conditioner and save a few hundred dollars.

Remember that you'll be dealing with electricity and high-pressure refrigerant, so if you're not comfortable handling either, proceed with care and utilize Phyxter to get in touch with a local professional.

As a certified journeyman in air conditioning, I worked for years on cooling systems ranging in size from window units to 6,000-ton chillers.

I'll attempt to share my best advice with you based on that expertise so you can install and maintain your air conditioning system affordably and safely.

You will need to contact a contractor if you need to have anything linked to the refrigerant system fixed, since I will only be discussing basic cooling system malfunctions.

Why isn't my house being cooled by my AC?

First things first, make sure the problem isn't with your furnace.

First things first: make sure your furnace (or air handler) and thermostat are operating properly.

To ensure that the A/C is "callie temperature change temperature, first set your thermostat to cold and lower the inside temperature.

After a few minutes, observe whether the furnace's fan is operating and circulating air around your house.

whether the furnace's fan isn't operating, turn off the furnace at the breaker box, wait a few minutes, then turn the breaker back on and watch to see whether the blower fan operates for an additional five minutes.

You should read another assistance article if the fan isn't turning on, which indicates that there may be an issue with the thermostat or the furnace.

If it does turn on, you can be certain that your air conditioner is the source of the issue.

Caution: Turn Off the Power

Recall that you should switch off the electricity before opening any of the panels on your air conditioner or furnace!

My air conditioner isn't working. How Can I Determine What's Wrong?

Central air conditioners may malfunction for a variety of reasons, so I'll guide you through my troubleshooting procedure, beginning with the most frequent reasons and working your way up to some more complicated but manageable problems.

- ☐ Reason 1: Plugged air filters or evaporator coil
- ☐ Reason 2: Lack of return air
- ☐ Reason 3: Plugged condenser coil
- ☐ Reason 4: Failed condenser fan
- ☐ Reason 5: Failed capacitor
- ☐ Reason 6: Failed contactor

Part Numbers and Names for Air Conditioners

You will need to know the general appearance and locations of the primary components before we go into each of the aforementioned "Reasons."

It's not always clear-cut, so you need to know which parts are most crucial to your cooling system and often cost the most to repair.

Begin debugging your own DIY air conditioner repair.

Reason 1: Evaporator coils or plugged air filters

A blocked filter must be the reason for 90% of repair calls for furnaces and air conditioners.

The likelihood that your dogs or cats are the problem is almost certainly 99%. During the cooling season, this is the main reason we get service calls!

The effectiveness of your system is greatly decreased when the airflow through the AC is restricted due to a blocked filter.

How to Replace the Air Filter in Your AC or Furnace

The procedure is turning off the furnace or air handler's electricity, locating the air filter, pulling it out,buying a new filter that has the same size and rating (the old filter will have the necessary information written on it).

Make sure the arrow written on the filter is pointing in the correct direction when you change it. It should be pointed in the direction that the air is moving, which is often towards the furnace. Don't trust the previous installation method.

You may resume the AC testing process by turning the furnace power back on and starting from the beginning of the book after installing the air filter.

Your airflow problem could have been resolved, but there might still be a lingering problem that needs time to resolve.

Methods for Restoring Frozen AC Evaporator Coils

The simplest method is to wait and let it thaw out, although it's not the most convenient.

Regretfully, the most probable cause of the freezing was a stopped air filter. Without air to absorb heat, the coil's temperature continued to decrease until it solidified into a block of ice.

Before you resume your air conditioner, it must be fully thawed; otherwise, you run the chance of it freezing up again.

One of the following may be done to expedite the process: configure your thermostat to turn off the air conditioning system but leave the fan running.

In doing so, the furnace fan will continue to operate and warm air will be blown over the coil.

Of course, you could also attempt the tried-and-true hairdryer approach, but you could simply run your fan instead of standing there for an hour.

You could increase the temperature and put your thermostat on heat. Moou really wanted to hurry things up.However, because your heat exchanger won't be receiving enough airflow, problems with furnace longevity may arise, thus this isn't advised.

Reason 2: Lack of Return Air

The reason this one might be hard to solve at the core is because it's really hiding from you.

The interior part of your HVAC system, the evaporator coil, is often found above your furnace. It requires airflow to absorb heat from the air; without it, the coil would freeze up and cease to function.

You tested the filter first, and it's OK. What should you do next?

There are two kinds of vents in your forced-air system: return air vents and discharge vents.

I'll go over the return air vents for our troubleshooting reasons.

They are often dirtier and bigger than the discharge vents.

They draw air into the system, while the discharge vents expel it after the air has been filtered at the furnace, making them dirtier.

Make sure that none of the vents—whether they are covered by dust or filth, obstructed by a door, or both—are covered or blocked.

Because the airflow in and out of your system was intended to be precisely calibrated, they must be clear.

Your air conditioner and furnace won't function properly and may even break if they don't get that amount of air.

Reason 3: Plugged Condenser Coils

You're not only coming close to solving the issue with your air conditioning system if you've made it this far, but you're also guaranteeing that it will operate more effectively in the future.

The largest energy user in your house is your air conditioning system, thus you will now be saving money.

Alright, let's test whether the coils on your condenser are clogged.

It's unclean and in need of cleaning, so let's take care of it right now if you haven't cleaned it as the image above shows.

Be careful: Verify that the outside condensing unit's electricity is off.

You may grab your water hose or compressed air now that the power is off. I like water since it's simpler and safer.

It is necessary to aim the hose directly at the outdoor coil, not at an angle, when you water it down.

Bending the coil fins will do more harm than good.

You may resume testing from the beginning of the book when the AC condenser coils are clean.

The compressor will be audible and the outside fan will operate if the system begins properly. The compressor has a sound similar to that of a noisy refrigerator; if it sounds louder, there may be an issue.

Another option is to wait and observe how your discharge air vents feel. If you have a temperature sensor, the air should feel around 20 degrees colder out of the vents than it does entering your return air vents.

The fourth step, a failing condenser fan, will need to be addressed if your air conditioner is still not operating.

Reason 4: Failed Condenser Fan

Although it could be thought of as easy to figure out, the repair is more involved than the other air conditioning repairs.This will need you to get out your tools.

Let's check to see whether the condenser fan has failed first.

Subsequently, ensure that your furnace and outside unit are powered back on, that the temperature is lowered, and that the thermostat is set to cool. Finally, verify that the outdoor unit is receiving the control signal.

There are two methods to do this: first, remove the condensing unit's control panel (which you may need to do otherwise), and then use a multimeter to make sure that the contactor is being "pulled in." The contractor will have around 24 volts across it.

Assume for the moment that the condenser fan is off and the contactor is operational.

If so, the compressor will turn on and off many times until the control board detects an issue and shuts the device out.

Alright, time to swap out the condenser fan motor.

As can be seen above, you must CAUTIONARY switch off the electricity to the outside condenser unit at the breaker.

Then, to remove the unit's cover, you'll need your nut drivers or screwdrivers. After that, you'll need to install the new motor using the old fan blade (unless the new motor came with one) and detach the motor from the cover.

In addition, I usually advise changing the fan blade as it might be difficult to remove the old shaft without the proper equipment.

Reconnecting the power lines to the new fan motor should be done properly according to the wiring schematic. You shouldn't be doing this step if you have trouble understanding a wiring schematic.

Let's now try turning on the unit's power once again to see whether it begins.

Fantastic! The fan is operating!

It's still operational.It will not quit operating!

Let's move on to the next possible problem. We'll examine replacing the capacitor next.

Reason 5: Failed Capacitor

According to the description above and the video lesson, the contactor is pulled in; nevertheless, because the compressor and condenser fan are not operating, it is possible that an AC capacitor has blown.

The steps to replace the capacitor are as follows.

How a dual start/run capacitor gets discharged

After removing the capacitor from the retaining bracket, contact the COMMON (or "C") and HERM (or "H") terminals with an insulated screwdriver.

Between the "C" terminal and the FAN (or "F") terminal, do the same. All that's required for single-mode capacitors is to provide a short between the two terminals.

How to Change a Dual Run/Start Capacitor

There is a capacitor in every air conditioning servicing unit.

If you're wondering what a capacitor accomplishes, it is used to improve the power of both motors during compressor and condenser fan starting by storing and releasing energy.

Additionally, it aids in mitigating grid voltage fluctuations, preventing damage to the motors.

The compressor and condenser fan motors have to work harder than intended as a result of capacitor failure over time.

To replace them, disconnect the old capacitor, snap a picture of its wires, and then put in the new one.

This is when your needle-nose pliers and insulated screwdriver come in handy.

Be careful: Before removing the capacitor, be sure it has been discharged.

Even after being disconnected from power, capacitors retain a significant amount of electric charge.

Reason 6: Failed Contactor

How to Change an AC Contactor That Has Failed

Let us carry out this process in detail.

First things first, confirm that the power is off at the breaker and use your multimeter to confirm that it is off.

Once the power has been turned off, detach the cables, install the new contractor, and then rejoin the cables. First, take some pictures of the wires that are connected to the contractor.

After making sure that every cable is connected properly, turn the power back on to test your air conditioner.

The 24 volt coil, which is what draws the contacts in and allows electricity to pass through the contactor, may be tested to see if your contactor is broken.

In essence, the AC condensing unit contactor coil receives a signal from the thermostat. The coil then energizes, produces a magnetic field, pulls the contactor down to complete the circuit, and allows 240 volts to pass through to the compressor and condenser fan.

In the event that the coil is receiving 24 volts but isn't being drawn down, the coil is defective and the contactor has to be changed.

An ohmmeter may also be used to verify continuity (that the 24 volt coil is continuous) in the coil.

The coil has to be changed if the ohmmeter reads open, often known as infinite.

Remember to contact your trustworthy local HVAC contracting business to come out and repair anything if you're ever uncomfortable testing the electricity—and you should be uncomfortable!

BOOK IV: ADVANCED HVAC TECHNIQUES

Introduction

Greetings from the world of Advanced HVAC Techniques, where sophisticated temperature management becomes a reality and conventional knowledge is pushed to the limit. This chapter takes you on a tour of cutting-edge ideas and specialized knowledge that will revolutionize the way you think about HVAC systems.

As we go through this chapter, picture a future in which energy efficiency is elevated to the status of art, troubleshooting becomes precise science, and the subtle differences between residential and commercial systems are analyzed to gain comparative understanding. Come explore with us the routes that lead to the highest level of professionalism in the HVAC industry—not simply competence.This is the chapter when mastery transcends the commonplace and theory becomes expertise. Welcome to Advanced HVAC Techniques, where you may start your path to become an HVAC virtuoso and where innovation meets application.

Chapter 1: Strategies for Energy Conservation

The following are five typical energy-saving techniques for heating, ventilation, and air conditioning:

- Chilled water supply temperature reset.
- Chilled water pumping differential pressure reset.
- Air handling unit supplies air temperature reset.
- AHU static pressure reset.
- Unoccupied HVAC setback.
- CHWST reset control strategy

Depending on the building geometry, load profile and variety, duct layout, and fan efficiencies, lowering the AHU duct SP setpoint may have a substantial effect on the energy consumption of airside fans and account for up to 5% to 7% of the HVAC power consumption of a building.

In order to identify the terminal box with the highest damper position, this technique necessitates polling each of the terminal boxes that an AHU serves on a time basis that may be adjusted. Setting all of the terminal boxes fed by the AHU to minimum cooling cubic feet per minute and determining the SP at the transmitter location that results in a terminal box with a damper position of 95% open, as well as setting all terminal boxes to full cooling cubic feet per minute and determining the SP setpoint that results in the terminal box with a damper position of 95% open, are how the minimum and maximum duct SP setpoints are determined during the test and balance of the air handling system. The polled terminal boxes' damper position is reduced by 0.1 inch of water gauge to determine the SP setpoint if it is less than 90% open. On the other hand, the SP setpoint is raised by 0.1 inch WG if the polled terminal boxes' damper position is more than 95% open.

Unoccupied HVAC setback

A popular and somewhat simple tactic is to adjust the HVAC setpoints when the building or its areas are empty. Typical reset parameters are as follows:

- Space temperature.
- Minimum zone airflow.
- Outside air quantity.

When these three tactics are used in tandem to cut down on cooling/heating energy as well as fan energy, maximum savings are realized. Assuming exhaust fan systems stay on during vacant times, care should be made to make sure that the overall building pressurization is not jeopardized while total and outside air volumes are setback.

These systems typically start based on a timetable for the time of day and include an override option that may be used with an occupancy sensor or a human push button.

Control strategy relationships

When using many techniques on a cooling system, it is essential to comprehend the links between the strategies that are being given. The easiest way to show the links between them is to divide them into waterside and airside categories. AHU supply air temperature reset and AHU SP reset are among the airside tactics, while CHWST reset and chilled water pumping differential pressure reset are among the waterside ones.

Because both waterside techniques make use of cooling load (the location of the chilled water valve), their interaction is significant. If both tactics were used at the same time, they would compete with one another and lead to a situation where an increase in CHWST would also cause the differential pressure setpoint to rise. The AHU coils will naturally need to adjust the differential pressure setpoint in order to maintain their respective discharge air temperature setpoints as the chilled water temperature rises. This is due to the fact that both solutions need monitoring the chilled water valve locations. Consequently, it is best to use the two procedures sequentially rather than simultaneously.

Greater than the energy reduction associated with chilled water pump DP reset and constant chiller discharge water temperature is the energy reduction (measured in kilowatts) associated with the chiller at increasing supply temperature setpoints, including the increase in pump kilowatts due to higher flow rate requirements. Therefore, the CHWST reset ought to happen first. The chilled water differential pressure may be changed to enjoy the extra energy savings associated with decreased pump horsepower after the CHWST reaches its maximum setpoint.

When the parameter used to alter it is examined, it becomes easy to understand the link between supply air temperature reset and SP reset. Both systems employ zone cooling demand to calculate the right setpoint value, but they calculate the necessary reaction using different control outputs. The cooling loop output of the terminal box is used to calculate the zone cooling requirement in the event of supply air temperature.

When an AHU SP reset occurs, the terminal box's damper position is used to calculate the zone cooling demand. Both tactics may be utilized concurrently since they rely on distinct characteristics to determine the response. Because the airside tactics will save more energy than the waterside strategies, this becomes even more crucial. This is due to the fact that airside systems account for a higher portion of a facility's overall energy consumption than waterside systems do.

Chapter 2: Proficient Troubleshooting Techniques

1. Pay attention to sounds and look for damage.

Typically, the HVAC system will find a method to notify you if anything is wrong. Occasionally, it produces strange scents, makes loud sounds, or the system doesn't heat or cool down efficiently. Make a note and do some investigation if you see any of these issues. You can usually find out why the system is creating that noise or scent on the internet.

First, go over any drafts. Your utility rates may increase due to an inefficient system brought on by a leaking duct. If the ductwork is severely damaged or too old, get an expert; however, you may seal tiny holes and cracks on your own. Drafts may also be reduced by cleaning the ducts and vents.

2. Examine the air filters

The simplest and most probable fix if you start to notice problems in your house is to change the air filters. Clogged, worn-out, or unclean air filters are often directly linked to an inefficient HVAC system. When was the last time you changed them, do you recall?

Debris may clog up vital HVAC system components and increase energy costs due to dirty air filters. When the filters in your HVAC system are clogged with dirt and debris, the system will not chill your house as effectively. This obstruction wears down the HVAC system and significantly raises the energy required to operate it. These elements may all significantly shorten the lifespan of your HVAC system.

The air filters may also be to fault if the air conditioner has problems keeping the temperature consistent. Blockage lowers the quality of the air inside your home by making your air conditioner work harder and enabling contaminants to pass past the filter. Replace your air filter on a regular basis to save your health and your unit.

Your filters should be changed every 90 days, but if they start to show signs of wear and tear and aren't functioning properly, you may change them every month. A fresh air filter fixes a lot of frequent problems, saves money, and enhances air quality.

3. Dust registers, ducts, and vents

Particles such as dust and debris are trapped in the registers, ducts, and vents when the air conditioner runs. The quality of the air within your home may suffer as a result. Maintaining clean ducts, registers, and vents greatly improves the efficiency of your air conditioner. We advise hiring a qualified specialist to clean your vents and ducts.

4. Check the thermostat.

You may be surprised to learn that the thermostat is more often the issue. Replace the batteries first, and then look for any loose connections. The thermostat should be the first item you examine if the air conditioning system doesn't turn on.

A functioning HVAC system is ensured by knowing if the thermostat is set appropriately. In the event that your system fails to start or starts slowly, make sure the thermostat is turned on.

Keep an eye on your thermostat to see whether it's the source of your HVAC issues. If the temperature inside is different from what the thermostat indicates, there may be a problem with your device.

5. Examine the External Unit

The exterior unit is the hub for airflow throughout your house, so although it may not be the first thing that comes to mind when you think of HVAC problems, it is a crucial component of your system. Your HVAC system is often located outdoors, and many types of debris may impair the cooling and air quality of the system.

Branches and other debris may get lodged within the device. Maintaining the unit's internal and external cleanliness will increase its efficiency.

To maintain the best possible performance from the HVAC system, open the condenser unit and remove any debris. Additionally, we advise trimming any trees close to the outside unit and getting rid of any overgrowth.

6. Examine the breakers box

Sometimes, after a storm, you may not have recognized that you lost electricity. You can't operate your air conditioner without electricity. Make sure the machine is powered on by quickly checking the break box.

The problem could be resolved by just flipping a switch. To make sure there are no electrical issues, you must reset the circuit breaker if it has tripped.

To prevent harm to yourself or your property, consult a professional to handle electrical issues since they may be dangerous.

7. Minimize Your Utility Bill

Higher utility bills are often associated with living in a more harsh environment. Your HVAC system may be too old to function well or may have sustained damage as a result of a sudden or steady increase in electricity expenditures.

By not adjusting your thermostat often, cleaning your air filters, and seeking periodic maintenance, you may reduce your power expenditures. All year long, 68–70 degrees is what we advise.

Chapter 3: A Comparative Analysis of Commercial and Residential Systems

HVAC systems for homes and businesses vary in terms of size, functionality, design, and optimal placement. As Superior Air explains the key distinctions between these two kinds of systems, continue reading.

Size

Compared to a typical commercial HVAC system, residential HVAC systems are smaller since they are intended for small- to medium-sized areas. Conversely, commercial HVAC systems are best suited for bigger buildings or structures. This clarifies the differences between their compressors, evaporators, thermostats, and other equipment and those used in residential settings.

Modular Design

Commercial HVAC systems use precise sensors, measuring tools, and adjustments to manage a variety of heating and cooling situations. This kind of performance calls for an intricate modular system architecture. Later on, the unit's capacity may be increased by adding more components to the unit.

Drainage

The amount of moisture produced by the two HVAC systems differs substantially. Commercial technology employs bespoke plumbing to remove more moisture than residential equipment, which merely uses a tiny pan and tube.

Location

Commercial HVAC systems are often positioned above a building to save space and avoid getting in the way of activities below. Condensers for residential HVAC systems are often positioned in the ground for simpler upkeep and cleaning.

Maintenance

Because residential HVAC systems are comparable in terms of design and specifications, they may be maintained using a uniform procedure. To diagnose and solve problems with commercial systems, however, experts that specialize in complicated modular modules are needed.

Cost

There is a significant cost difference between HVAC systems for homes and businesses. Taking into account all we've covered so far, commercial units are pricey due to their application-specific design. Furthermore, since installing, maintaining, and repairing commercial HVAC systems takes more time, a commercial HVAC installer bills at a higher rate.

Chapter 4: Pathways to Becoming a Professional in the HVAC Field

A career in heating, ventilation, and air conditioning (HVAC) combines scientific knowledge, practical experience, and a dedication to providing comfort in a variety of settings. It is a fulfilling path. Whether your career goals are to be an engineer, expert, or HVAC technician, getting to the professional level requires a mix of training, real-world experience, and continuous improvement. We'll go over all the stages and things to think about in this in-depth guide to help you become a seasoned HVAC specialist.

1. Foundations of Education: Establishing the Framework

Seek Out Formal Instruction

Get a strong educational foundation first. Think about signing up for a respectable HVAC course at a community college, technical school, or vocational training facility. Fundamental concepts, system design, installation methods, and troubleshooting abilities are usually covered in these courses.

Obtain a Degree

A degree in HVAC engineering, or a similar discipline, may provide a better comprehension of the theoretical elements, system design, and energy efficiency concepts for people looking for advanced employment. Having an associate's or bachelor's degree improves your resume and gives you access to a wide range of job options in the HVAC sector.

2. Practical Training: Connecting Theory and Application

Apprenticeships

Have a look at an apprenticeship program; it's a great way to get practical experience and collaborate with seasoned experts. This hands-on training gives you a comprehensive grasp of HVAC systems, exposes you to real-world settings, and improves your practical abilities.

Practical Experience

A lot of people in the HVAC industry start out as entry-level technicians and learn on the job. This hands-on learning opportunity is priceless as it provides insights into the subtleties of various tools, systems, and client interactions. Take advantage of the chance to work with knowledgeable mentors on installations, maintenance, and repairs.

3. Certifications: Proof of Expertise

EPA Accreditation

Obtain certification from the Environmental Protection Agency (EPA), which is necessary in order to handle refrigerants. This accreditation demonstrates your dedication to regulatory compliance and environmental safety.

Certifications for the Industry

Go for industry-recognized credentials, including HVAC Excellence or North American Technician Excellence (NATE). These certificates increase your employability and reputation by attesting to your expertise in certain HVAC fields.

4. Specialty: Developing Your Own Market

Investigate Specialized Domains

As you have more expertise, think about focusing on certain HVAC disciplines like energy management, commercial systems, or industrial refrigeration. Gaining specialization may open up more advanced positions and boost the demand for your knowledge.

Ongoing Education

Keeping up with the latest advancements in HVAC technology is crucial. To stay up to date with developments in the sector, attend conferences, seminars, and workshops. To expand your knowledge and skill set, think about taking on further courses or obtaining higher certifications.

5. Career Ladder Climbing: Professional Development

Pursue Professional Growth

After gaining more experience and credentials, look at possibilities for professional progression. Positions like project manager, systems engineer, or HVAC supervisor can fit your goals.

Become a Member of Professional Associations

Join HVAC trade associations like the American Society of Heating, Refrigerating, and Air-Conditioning Engineers (ASHRAE) or the Air Conditioning Contractors of America (ACCA). Making connections with business leaders in your field might open up job chances, mentoring, and insightful perspectives.

6. Standards of Ethic and Customer Service: Establishing a Credibility

Respect Ethical Principles

Uphold a dedication to moral behavior, openness, and honesty in all facets of your professional life. Establishing a reputation for reliability is essential in the HVAC sector.

Be the best in customer service

Develop your interpersonal and communication abilities. In addition to improving your professional reputation, offering top-notch customer service also fosters client loyalty and contentment.

In the HVAC industry, being a professional is a dynamic and continuous process. You may have a successful and rewarding career in this important business by combining education, practical experience, certifications, expertise, and a dedication to ethical standards. As you advance on your route, welcome the always changing HVAC technology and never stop striving for excellence in the efficiency and comfort of the people you serve.

BOOK V: ESSENTIAL TOOLS FOR HVAC

Chapter 1: Essential HVAC Tools and Equipment

Basic HVAC Tools

1. Screwdriver

A screwdriver is one of the most important tools for the HVAC industry as well as other industries and areas. Drives with Phillips and Flat heads are two common varieties. However, other screwdriver kinds—like Torx, Hex, Square, etc.—are also often used in HVAC systems.

Screwdrivers are often used by technicians as pry bars, chisels, and scrapers. They shatter quite readily as a consequence. Screwdrivers are a useful tool for many years in the HVAC industry, but we advise you to use them exclusively for the specified uses.

2. Hammer

A hammer is another common hand instrument that every HVAC professional has to own. Nail hammers, ball peen hammers, and mallets (or other soft hammers) are the three most popular kinds of hammers.

Hammers may be used to strike an item with great power or to drive a nail. You may need a lightweight hammer or a standard hammer, depending on your needs.

3. Plier

A plier is a highly helpful hand tool that you may use to twist objects, cut wires, and reach into holes. General Purpose Pliers (Linesman Pliers), Cutting Pliers, Needle-Nose Pliers, Curved Needle-Nose Pliers, Tongue & Groove Pliers, and Locking Pliers are a few typical varieties of pliers.

In their toolkit, HVAC technicians need to have a minimum of two varieties of pliers.

4. Wrenches

TornelsThe professional working on HVAC installation or maintenance may encounter several pipes, nuts, and bolts. You will need a variety of wrenches to tighten or loosen them.

One of the most important equipment for any HVAC project is a pipe wrench. The technician can grasp pipes and other similar circular surfaces with ease because of its distinctive form and teethed head. Pipe wrenches in sizes of 8, 10, and 12 inches are widely available.

Additionally useful are combination wrenches, box end wrenches, and open end wrenches. You may use a single-sized nut or fastener with each of these wrenches, so make sure you have a set of them.

Another well-liked kind of wrench is the adjustable wrench. You can work on fasteners of various sizes with only one tool. Furthermore, extremely helpful is a Ratchet Wrench. A nut or fastener may be tightened (or loosened) without removing or adjusting the tool.

5. Socket Set

A complete collection of tools, such as socket wrenches, reversible ratcheting handles, socket extensions, universal joint sockets, etc., are often included when you purchase a socket set.

There are many sizes of socket wrenches, ranging from 1/4-inch to 1-inch. The technician can access deep or awkwardly angled nuts and bolts with the use of a socket wrench extension.

6. Allen Keys, or Hex Keys

Hex keys, also known as Allen keys, are six-sided instruments that let you operate on slots with a hexagonal form. Hex keys are often used by technicians to install or remove blowers from motor shafts.

In addition to standard right-angled Allen keys, Hex keys with T-handles are also available.

7. Drills

An HVAC technician may find great use in an electric drill. To run cables or fasten an item, drill holes and screwdrivers may be used. Although there are corded and cordless electric drills available, we suggest a good cordless drill, ideally with a hammering motion.

A corded drill will be a more cost-effective and powerful choice if you are certain that you will always have access to electricity.

8. Reciprocating Saws

It is necessary for HVAC technicians to cut through pipes, plywood, and drywall. A Sawzall, often known as a reciprocating saw, is a common instrument for all these cutting tasks.

Similar to electric drills, corded and cordless reciprocating saws are offered. Generally speaking, corded reciprocating saws are much less expensive than cordless reciprocating saws. Batteries and a charger may not always be included with cordless devices; these items must be bought separately.

When choosing the ideal reciprocating saw that meets your needs and budget, do some study. Even while it could seem like a big investment, in the long term, it will save you a ton of time and work.

9. Hacksaws

An easy-to-use hacksaw is the ideal instrument for cutting thin metal sheets fast. Conduits and general-purpose pipes (not refrigerant-carrying ones) may be cut in addition to metal sheets.

10. Chisels

It is not appropriate to use a screwdriver as a chisel. To remove extra material or for comparable uses, get a chisel. One end of a cold chisel is sharp, while the other is a hit head. They may be used to cut metal that is somewhat softer than the chisel itself, and they are often constructed of steel.

A Brick Chisel is another kind of chisel that's well-liked by HVAC workers. These chisels are for cutting or trimming bricks, as the name implies.

11. Files

You need a file to smooth or remove any sharp edges from metal (or any other material). They are available in an array of shapes and sizes.

Common file formats are Wood Rasp Files (Flat, Half-Round, Round), Round files, Half-Round files, and Double Cut or Curved Files.

12. Metal Shears

Shears are a highly helpful instrument for cutting metal sheets into long, straight pieces. HVAC professionals seldom have to handle duct installation tasks like this, but sometimes they must drill holes in the ducting in various orientations.

13. Cutters

Air Conditioning Technicians use a variety of tubes and pipes in their job. Part of the task is to cut them so they fit neatly and properly.

The three most popular kinds of cutters that you may use to cut copper pipes, PVC tubes, and other materials are tube cutters, pipe cutters, and metal cutters.

14. Tin Snip

Tin Snips are helpful for making little cuts, whereas Metal Shears are great for larger cuts. You can cut metal straight using Line Type Tin Snips. There are also Tin Snips unique to the Left and Right directions.

There is a Curve Type version of Tin Snips also. Curve Types are also offered as Left-Curve and Right-Curve Tin Snips, much like Line Types.

15. Hand Seamer

When dealing with sheet metal, you may sometimes need to flatten or bend it at various angles. For tasks like these, a hand seamer is the ideal instrument. HVAC technicians can bend sheet metal at precise and tight angles by using a hand seamer.

Hand seamers may be used to smooth out creases and bends on sheet metal in order to remove any undesired shapes. They come in a variety of jaw widths, from three to nine inches.

16. Folding Bar

A Folding Bar or Drive Bar is an additional tool that is useful for working with sheet metal. The sheet metal may be "folded" at any desired angle, or often at a 90° angle, by fastening the Folding Bar to the vice.

17. Awls

A long, sharp tool called an awl is helpful for perforating holes in sheet metal. Awls are also used by HVAC professionals to mark surfaces, such as score lines. There are several sizes of awls available.

18. Pipe Crimpers

HVAC technicians may crimp copper and PEX pipes using metal crimpers (also known as pipe and tube crimpers). These Press Tools can create watertight seals without the need for welding or soldering. Pipe cutters are quite common in both conventional plumbing and HVAC applications.

19. Wire Strippers and Crimpers

HVAC technicians have to deal with wires of varying gauges while making various electrical connections. A wire stripper is a highly useful equipment because it makes it simple to remove the outer layer of plastic or rubber from wires so that electricity may connect.

Another helpful equipment for crimping wire ends is a wire crimper. This enables the technician to finish the wires correctly.

HVAC Specialty Tools

1. Multimeters

A multimeter is an essential equipment for every tradesperson, whether it an electrician or an HVAC technician. Although digital multimeters, or DMMs, come in a range of pricing ranges, we advise you to spend your money on an excellent HVAC multimeter.

You may examine the conductivity and voltages of the switches and cables. Many functions, including temperature measurement and live voltage detection, are available in modern multimeters.

2. Warmth gauges

Quality Thermometers are essential for HVAC workers to get precise temperature readings. These are quite helpful during service since, in the event that there is insufficient cooling, they may also aid in adjusting the refrigerant level.

Different sizes and forms are available for thermometers and temperature sensors. To place the thermometers into the ducting, certain models come with pointed probes. We advise you to get temperature sensors and digital thermometers.

3. Suction Pumps

A powerful tool for service or repairs, an HVAC vacuum pump may remove moisture or vapor from the device. Reclaimer pumps, which enable the recovery and subsequent use of refrigerant fluid, are another tool used by HVAC specialists.

4. Filter Pullers

The technician must reach filters that are particularly challenging to access while maintaining or repairing HVAC units. It might be difficult to remove filters from such difficult locations. Filter pullers are specialized instruments designed to grasp and remove filters using a telescopic handle.

5. Gauges for Refrigeration

An HVAC expert can measure the refrigerant pressure precisely with the use of a refrigeration gauge. As the pressure decreases, these gauges assist in determining if there is a leak in the line.

6. Gauges Manifolds

This HVAC tool is crucial. Manifold gauges aid in the accurate diagnosis of high and low pressure in air conditioning systems by technicians. Manifold gauges, such as those for R-134A, R-12, and other types of refrigerant, are available.

High-quality digital gauges that provide precise pressure measurements are now readily accessible. However, analog gauges from the past are still widely used.

7. Hose Gauges

Unfortunately, refrigeration gauges and manifold gauges do not come with the hoses that are required for the HVAC technician to monitor the refrigerant pressure.

Gauge hoses with Quick Connect connections are a common kind. These gauge hoses' construction ensures that there won't be any leaks outside the air conditioner.

8. Tool Bags for HVAC

A number of tools must be carried by an HVAC professional while doing installation, maintenance, or repair work. You may arrange all the equipment in their compartments or pockets in a tidy and efficient manner with an appropriate HVAC tool bag, which will save you a great deal of trouble.

There are many different alternatives available if you search around for high-quality HVAC tool bags, including as backpacks, waist bags, and hand-held models.

Advanced Instruments for HVAC

1. Core Removal Instruments

An essential component of the Schrader valve is the valve core, which is a poppet valve that is spring-actuated. To change and remove the Schrader cores while fixing them, you'll need specialized tools.

Numerous Core Removal Tools include a unique torque mechanism. The torque mechanism clicks when you tighten the valve core, indicating that the thread is completely tight and not causing any leaks.

2. Rehabilitation Centers

HVAC professionals have a few rules to observe while handling refrigerants. It is strictly forbidden by law to release refrigerants into the environment. A simple piece of equipment called a recovery unit aids the technician in extracting all of the refrigerant from the air conditioning system.

Both liquid and vapor refrigerants may be recovered using refrigerant recovery units, often known as tanks, machines, or units.

3. Fin straighteners with coils

An air conditioning system's condenser becomes clogged and dusty over time. Both the evaporator and the condenser may experience this. Fins on coils may wrinkle as well.

A simple instrument exists in the shape of a coil fin straightener. The HVAC expert may clear trash and even straighten the coil fins with the use of these metal or plastic comb-like tools.

4. Paintball Guns

One of the most important aspects of HVAC installation and maintenance is building leak-free ducting. HVAC technicians may stop air leaks by using Caulk Guns, which are extremely easy to use and practical tools for sealing small gaps and openings.

The majority of caulk guns are manually operated instruments that just need the pull of a trigger. By doing this, you may squeeze the cartridge and get the caulk out. To prevent mess, choose for a caulk gun with a dripless function.

5. Regulators of Nitrogen

robust In HVAC systems, nitrogen regulators are used for leakage checks, pressure build-up, and purging. These leak checks are carried out by HVAC technicians while the system is devoid of refrigerant.

When adding refrigerant to the air conditioning system, nitrogen purging is an essential step. Any oxygen in the system is displaced by the nitrogen gas, which also reduces the amount of oxides that develop. You need high-quality nitrogen regulators to regulate the nitrogen gas's pressure.

6. Detectors of Carbon Monoxide

When constructing, maintaining, or fixing HVAC systems, experts need to make sure there is no carbon monoxide present in the building. For this, carbon monoxide detectors are very helpful.

In a house or place of business, carbon monoxide may come from a variety of sources. These consist of furnaces, generators, space heaters, fireplaces, and so on. Before beginning any repair or service, an HVAC technician's first responsibility is to utilize an appropriate CO (carbon monoxide) detector and look for any leaks.

7. Leak Detectors for Refrigerators

The proper management and upkeep of refrigerant is essential to the proper functioning of an air conditioning system. Refrigerant leaks may be caused by temperature, vibrations, poor joint fittings, incorrect welding sites, and hostile conditions. The air conditioning system loses a significant amount of efficiency if these leaks are not found and fixed.

Refrigerant leak detectors that are electronic are very user-friendly and capable of precisely finding even tiny leaks. In addition, there are several kinds of leak detectors, including corona suppression, heated diode, ultrasonic, and infrared.

8. Cameras with Thermal Imaging

When dealing with HVAC systems, thermal imaging cameras make it easier to locate issues quickly. Once an HVAC technician has resolved them, they can demonstrate to the client that the issue has been resolved.

Due to their exorbitant cost, many HVAC professionals were previously hesitant to make the investment in appropriate thermal imaging gear. The price of a good thermal camera has decreased dramatically in recent years.

Thermal imaging cameras provide a number of advantages. They swiftly assist you in locating leaks, whether they be water, air, or even electrical. You may demonstrate to the client precisely what the issue was and how you resolved it.

9. Examiners of phases

HVAC technicians also need to be properly knowledgeable about mains AC electricity and the many instruments that go along with it. A Phase Tester is one such instrument. When fixing or maintaining an HVAC system, these electrical instruments come in handy for determining the HVAC unit's phase or current.

10. Psychrometers

A psychrometer is a specialized instrument used to determine the air's relative humidity. The HVAC system's primary responsibility in a closed room is to maintain appropriate moisture levels. HVAC technicians use psychrometers to measure relative humidity and the quantity of moisture in the air while doing repairs or maintenance.

The Sling-type Psychrometer is a widely used instrument. Two thermometers are inside; one is damp and the other is dry. These days, computerized ones with sensors and electronics are available.

11. Cleaning Agents

Any HVAC system's line may get clogged by debris, carbon residue, and moisture contamination, among other things. An extended period of use of the HVAC unit might exacerbate the issue.

The refrigerant line needs to be cleaned by HVAC technicians. You will require refrigerant flushing solvents to do this efficiently. Any impurities in the refrigerant lines will be dissolved by these solvents, which will also dissipate after the task is over.

12. Tanks for Storing Refrigerant

Technicians may securely store the recovered refrigerant in refrigerant storage tanks after removing it from the HVAC unit. There are several capabilities accessible for them. The option that best satisfies your needs must be chosen.

13. Scales for Refrigerants

The HVAC technician must replenish the system with the appropriate amount of refrigerant, regardless of whether there is a leak or the technician completely drained the refrigerant out of the unit. Refrigerant scales and charging scales are excellent tools for this.

The HVAC professional may precisely weigh the refrigerant in a cylinder before charging or refilling it by using a refrigerant scale.

HVAC Safety Equipment

1. Summaries

Air Conditioning When working on air conditioning systems, technicians need to dress appropriately. For difficult tasks involving a lot of sharp items or dust, coveralls and overalls are appropriate.

2. Gloves

Another crucial safety item to shield hands from chemicals, sharp objects, and even rough handling is a pair of hand gloves. Your choice of gloves should protect your hands while still being comfortable enough to handle tiny things like hand tools.

3. Eyewear

It is crucial to shield your eyes from dust, dirt, and other small particles. So, safety eyewear is necessary for any HVAC technician. They will protect your eyes from all of these tiny particles so you won't have to worry about damaging them while working.

4. Masks

Air Conditioning The right masks must be worn by technicians as part of their personal protective equipment. Air conditioning system repairs, servicing, and maintenance may expose you to a variety of debris, including dust, chemicals, and even residues in the lines.

You can keep microscopic pollutants out of your mouth and nose by wearing an appropriate industrial-grade mask.

5. Plugs for ears

Working with compressors, power tools, and other noisy engines and machinery is a constant part of the job description for HVAC technicians. Therefore, you must use appropriate industrial-grade earplugs or ear muffs to protect your ears.

6. Footwear

You can operate on a variety of terrains and keep your feet safe by wearing a pair of sturdy steel-toe work boots. Wearing appropriate footwear is essential whether working on the floor, climbing ladders, or even entering air ducts.

HVAC Attachments and Utilities

1. Levels

To ensure that everything is aligned correctly, you must use a level while installing HVAC units, plumbing, or electrical components. For marking on walls, basic liquid bubbles work well, although more sophisticated laser levels are available.

2. Measurements using Tape

HVAC specialists use tape measures to take precise measurements. A lengthy tape measure is essential for measuring ductwork sizes and installing HVAC units. Make sure the tape measure is at least 25 feet long. The longer the better when it comes to tape measures.

3. Lightsabers

HVAC technicians have to access areas that are poorly lit. Therefore, you may light up the area around the HVAC unit or within the ductwork using a suitable flashlight or headlamp. We advise getting a high-quality head-mount light so you can free up your hands.

Head-mounted lamps are helpful, but standard flashlights may also be used to aim light in any desired direction.

4. Stair Climbers

An HVAC technician must spend hours standing on step ladders. Step ladders are your best friend while working on ducting, electrical wiring, or refrigerant lines. Although there are many varieties of ladders, most chores may be completed with a simple step ladder.

5. Hand Tool Sets

When cutting through tiny objects like boxes, insulation, cables, etc., a utility knife comes in handy. Make sure you always have a few utility knives on hand, in various sizes, along with extra blades.

6. Power Cords

It is possible that power may not always be available. But electricity is required in order to operate certain power tools. Extension cords come in handy here to make your task easier. Reciprocating saws and electric drills are among the tools used by HVAC technicians. Thus, be sure to choose an Extension Cord of appropriate length and quality.

7. Squares

Triangle squares, often known as speed squares, are excellent tools for marking 45° and 90° in addition to measurements. Speed

Squares are a useful item to have in your toolkit as you never know when you'll need them.

8. Exam Mirrors

HVAC technicians may check and examine deeply into the HVAC unit with the use of telescopic inspection mirrors. To light up shadowy regions, certain inspection mirrors have LED lights built in.

9. Zip ties and cable ties

You have to work with wiring, cables, pipes, and other things as an HVAC technician. Use Zip Ties or Cable Ties to secure them together rather than dangling them in midair. Keep a bag of cable ties in your toolbox at all times.

10. Fixing Guns

When working on ductwork, you must adhere the foil lines to the duct's surface. For this kind of work, staple guns are a very easy and effective instrument.

11. Adhesive tapes

With duct tape, there's nothing you can't repair. Joking aside, duct tape is a useful tool for securing objects.

12. Trays with Magnets

You may save a ton of time and work by using a little Magnet Tray to keep all of your metal screws, nuts, tools, and other items. You don't need to be concerned about losing such little things.

13. Fasteners

An extra pair of screws is a necessary tool for HVAC technicians. It's possible that some screws are already missing or that you will lose some. Thus, be sure to choose a package of screws that includes all widely used sizes.

14. Fasteners

HVAC technicians sometimes do electrical installations, therefore they are not strictly electricians. Rerouting the cables or installing more wiring may be necessary for this. With the aid of wire nuts, you may connect fresh wire to existing ones without the need for soldering or other challenging tasks.

15. Pullers for fuses and fuses

Sometimes, a blown fuse might be the only issue with an HVAC unit. Thus, have an extra set of fuses with varying amperages on hand.

Make sure you also get a fuse puller since you never know how simple or challenging it will be to locate the fuse and remove it.

Chapter 2: Practical HVAC Tips and Tricks

HVAC technician tips for beginners

Prior to addressing an on-site HVAC customer request, a novice HVAC technician should be knowledgeable with several technicalities and fundamental best practices.

1. Make sure you're safe

As a novice in the HVAC sector, you must take care to stay away from risks and physical strain. According to the Bureau of Labor

Statistics, working in the HVAC industry is regarded as a potentially lethal profession.

For preventative purposes, you should make sure that you wear suitable clothing, drive safely, and have access to industrial-grade footwear.

2. Pay attention to the details

Pay attention to HVAC nuances that may improve the correctness and precision of the work while assisting you in providing a first-rate client experience. Keep an eye on the needs of your clients and adjust your HVAC service offerings accordingly.

During the solution delivery process, concentrate on communicating with the customers and management to reduce any misconceptions and future difficulties.

3. Provide a precise diagnosis

Make certain that you provide a precise and safe HVAC solution and that the homeowners feel protected. Before addressing the customer needs on the ground, have faith in your abilities and make the necessary knowledge and information acquisitions.

Examine the underlying source of the issue prior to trying HVAC system repair. Take your time in order to reduce the possibility of tragic mishaps.

Before making any changes to the current HVAC system, make sure you fully disclose your diagnosis to your customers to avoid any future misconceptions.

4. Make a help request

As a novice in the field, you need to seek the appropriate guidance from your more experienced colleagues to help you prevent mistakes or defects in your HVAC installation or repair job.

whether you're not sure whether the problem is a refrigerant leak or something more complicated, like a compressor or condenser blower, don't presume.

Get the expert opinion of your supervisor on the particular issue at hand and the proposed solution for the customer.

5. Communicate with the customer

Always remember to listen while interacting with clients. You can solve problems more quickly and skillfully by understanding client needs and interacting with them. This will help you meet your HVAC service objectives.

For you to be completely certain of your remedy and provide a first-rate client experience, you should find out the customer's viewpoint on the problem and what they want.

6. Maintain communication with your group

Maintain open lines of contact with the senior team and management to supply the required solutions and get around any dynamic challenges that may arise throughout the execution process.

To keep a strong connection with your superiors, you may also provide your team real-time information on the status of the HVAC project.

7. Maintain organization

Maintain organization in your company's operations to steer clear of various inefficiencies that lead to a bad client experience. Dealing with a technician that looks to have been hiding for a long time, leaves mud stains on the carpet, or is disorganized in their van is not something that anybody enjoys.

When leaving a project site, go above and above to provide excellent HVAC service and tidy up everything to return the area to how it was when you arrived.

8. Have a direct conversation with the customer

While the HVAC service is being performed, have a conversation with your customer and ask them specific questions. It helps you to investigate HVAC services more thoroughly and raise the quality of the service encounter.

Obtain the important information that was overlooked during the explanation of the customer's needs or the first inspection.

9. Initial examination

You need to carry out an initial assessment that may provide a more thorough understanding of the circumstances and the needs of the customer.

Examine the regions that need HVAC maintenance and verify other fundamental aspects to eliminate uncertainty and customize your HVAC repair offerings.

HVAC advice for customers

Before calling for professional heating, cooling, and ventilation services, homeowners should be aware of a few pointers in addition to the HVAC specialists.

10. Be familiar with HVAC systems

The HVAC system that is installed in the client's home or place of business should be known to them. It facilitates the HVAC company's ability to provide customized HVAC solutions without future misunderstandings or uncertainty.

Additionally, it may save needless rework and project costs, which can lead to efficient service operations.

11. Replace the air filters

By changing the air filters, you can guarantee that the HVAC system operates smoothly and prevents future problems.

To reduce the total cost of HVAC maintenance and enhance the performance of the HVAC system, homeowners or managers of commercial buildings may request that the HVAC contractor or service providers replace the filters.

12. Sanitize the condenser.

Due to its location in the side yard, the condenser is frequently forgotten by patrons. Shrubs and debris may accumulate around the unit, obstructing ventilation and causing other issues.

In order to chill the room, it may reduce the air conditioner's quality, which would not be good for the consumer. To clean the condenser unit, the homeowner should speak with many HVAC businesses or refer to HVAC professional suggestions.

13. Keep the vents open.

Several customers shut the roll vents to save energy, which reduces the HVAC system's efficiency and raises energy usage.

Making sure the air vents are open promotes even temperature dispersion throughout the house or business, which improves the efficiency of the heating and cooling system.

14. Dust the ducting

The HVAC system sometimes produces loud sounds and operates less effectively than usual. It is the responsibility of business space managers or homeowners to make sure that the duct is clear of debris and problems.

Professional HVAC technicians should be contacted to replace or clean the ducts, and the equipment should be inspected on a regular basis.

15. Employ experts

Installing HVAC systems should not be attempted by a homeowner who finds it difficult to handle technical HVAC activities.

Professionals who use contemporary tools and technology to provide excellent customer service should be given the task. It may reduce problems and difficulties while saving time and energy.

Summertime advice for HVAC technicians

Summertime demands extra attention for the HVAC system to prevent visible issues.

16. Make use of intelligent thermostats

If you operate in a warm climate such as Los Angeles, Palm Desert, or Orange County, you should inform your customers or potential leads about the benefits of installing a smart thermostat, which may help them save money and provide the best possible customer experience.

The goal of the smart equipment is to reduce the amount of work required to maintain conventional air conditioners while maximizing the effectiveness of the current business environment.

Tips for springtime HVAC maintenance

The springtime humidity puts more strain on the HVAC system, which might lead to problems. A simple step might help to reduce the problems.

17. Preventive dehumidifier maintenance

In humid climes, a dehumidifier may be the difference between life and death. The inconvenience and issues with the AC unit may be reduced by checking the dehumidifier.

The HVAC system's performance in the spring may be enhanced by finishing the debris removal and drying off the unit.

Wintertime advice from HVAC technicians

Before the HVAC system in the winter can warm up the room, it has to be in excellent working order.

18. Upkeep of furnaces

In order to prevent homeowners from having a very severe winter, regular furnace maintenance is essential. It is crucial to examine the furnace's finer points and make sure any dirt and debris have been cleared out.

Because of the increasing complexity of the process, only a professional team of HVAC professionals can provide high-quality furnace maintenance service.

Professional advice for HVAC technicians

A few key pointers will assist you as an established HVAC independent contractor or company owner to enhance your HVAC service operations.

19. Improve the experience for customers

Your position in the market may be strengthened and your service operations can be expanded to new heights by concentrating on enhancing your customers' total customer experience.

Positive word-of-mouth marketing may help you lower the cost of customer acquisition while increasing client retention. Before beginning service operations, ask your customers thorough questions that will provide a strong basis.

Make sure you have the skills, resources, and understanding necessary to complete the desired HVAC job. Don't make promises you can't keep that might damage your reputation as a brand in the marketplace and the consumer experience.

Assign the appropriate work to your team so that you may finish the project as quickly as possible while maintaining the quality of the client experience.

20. Mentor incoming HVAC technicians

The development of next HVAC leaders and experts should also be a priority. After hiring young, inexperienced HVAC technicians, you should provide them access to specialists who can expand their knowledge and skill set and on-the-job training.

To reduce the likelihood of errors and accidents, make sure HVAC newbies are not given complicated duties and are supervised by professionals.

For newcomers, offering top-notch practical knowledge and expertise may fortify your staff and assist you in growing your company to new heights.

21. Make use of specialized management software

Employing expert HVAC business management software may increase efficiency and reduce reliance on human labor while growing your company's size. Make sure you can automate client request handling and provide a high-quality customer experience.

To increase the productivity of your staff, you may streamline the procedure and reduce its complexity. Establish a channel of communication between your management team and field staff to help your HVAC technicians avoid the dynamic problems.

The program may assist you in creating bills, giving customers access to the HVAC technicians' real-time locations, and producing thoroughly examined performance data for optimization.

By using precise and accurate data to accomplish the required business goals and differentiate yourself from other HVAC service providers, you may enhance the performance of your current services.

VIDEO COURSE

(kindly type in those links into your browser)

Video Module 1: Basic HVAC Components
Go here: http://tinyurl.com/53a9849u

HVAC Parts and Functions
Go here: http://tinyurl.com/mvpr9ujr

Video Module 2: DIY HVAC Maintenance
Go here: http://tinyurl.com/vnue3bzy

Practical Demonstrations of Maintenance Procedures
Go here: http://tinyurl.com/yc77r349

Extra Resources for HVAC Enthusiasts

Recommended Books and Further Reading

"HVAC Systems Design Handbook" by Roger W. Haines and Michael E. Myers:

This thorough manual offers a thorough examination of HVAC systems by combining theory with real-world applications. For those who are just starting out and want to understand the basic ideas behind system design, this is a great resource.

"Modern Refrigeration and Air Conditioning" by Andrew D. Althouse, Carl H. Turnquist, Alfred F. Bracciano, and Daniel C. Bracciano:

This book, which is regarded as a classic in the industry, explains the fundamentals of air conditioning and refrigeration systems. It accommodates novices while providing a strong basis for more complex subjects with its concise explanations and informative illustrations.

"HVAC Fundamentals" by Samuel C. Sugarman:

This book, written with beginners in mind, is a useful reference that covers the fundamentals of HVAC (heating, ventilation, and air conditioning). It is user-friendly for novices, with examples and applications drawn from real-world situations.

"Audel HVAC Fundamentals, Volume 1: Heating Systems, Furnaces, and Boilers" by James E. Brumbaugh:

This book mostly addresses boilers, furnaces, and heating systems. It offers a practical method for comprehending these elements, which makes it a great tool for novices who are keen to learn about HVAC mechanics.

"Air Conditioning and Refrigeration Repair Made Easy" by Hooman Gohari:

This book, which is aimed for do-it-yourselfers, makes air conditioning and refrigeration system repairs easier. It enables novices to handle typical problems with step-by-step instructions and troubleshooting recommendations.

Online Forums and Communities

HVAC-Talk Forum (hvac-talk.com):

A thriving online forum where HVAC experts and fans converse about troubleshooting problems, offer useful advice, and share insights. Novices may get insightful guidance from seasoned participants.

Reddit HVAC Community (reddit.com/r/HVAC):

Beginners may engage in debates, exchange experiences, and ask questions with a worldwide network of HVAC experts, technicians, and hobbyists on the HVAC subreddit.

HVAC School Community (community.hvacrschool.com):

This platform, which is connected to the HVAC School podcast, provides discussion boards for a variety of HVAC subjects. The combined expertise of experts in the industry may help beginners.

Contractor Talk Forum (contractortalk.com):

This forum unites contractors from different trades, while it is not only HVAC-focused. Beginners in HVAC may learn more about the bigger construction sector and how HVAC works from a larger perspective.

DoItYourself.com HVAC Forum (doityourself.com/forum/hvac):

This site, which caters to do-it-yourselfers, offers a place where novices may ask questions about HVAC installations, repairs, and projects. Members of this encouraging community exchange problems and solutions.

Greetings, Reader

I hope you are doing well as I write this. We appreciate you taking the time to read The Updated HVAC for Beginners 2024. I value your opinions much and would be interested in knowing what you think of the book. Your viewpoint affects whether you have begun, are halfway through, or have done reading.

I really value your input, which will help me improve my next efforts.

We appreciate you taking the time to comment on The Updated HVAC for Beginners 2024. Your support means the world to me.

Enjoy your reading!

Conclusion

We say goodbye to this thorough guide with a feeling of success as we approach the last chapter of "The Updated HVAC for Beginners 2024" and with fresh information. Our trip has been nothing short of an adventure, from learning about installation procedures to exploring maintenance and troubleshooting to piecing together the historical background of HVAC.

We went over the foundations of HVAC on the pages that before this conclusion, understanding the complex language of parts, systems, and the science behind our temperature control equipment. We descended into the painstaking process of installation expertise, seeing systems come to life under our direction and experiencing the birth of comfort. Going forward, we revealed the mysteries of upkeep and troubleshooting, giving you the ability to take command of the health of your HVAC system and confidently address typical problems.

Since empowerment is based on knowledge, this book aims to provide you with the skills necessary to successfully navigate the complicated world of HVAC. This is only the beginning of your adventure; you may grow and hone your abilities to become a professional HVAC installer or a do-it-yourself enthusiast hoping to improve the comfort of your house.

We tried to achieve a harmonic balance between theory and practice throughout this tutorial. Every chapter sought to provide a thorough knowledge that goes beyond the surface, covering everything from the theoretical foundations of HVAC systems to practical installation methods. The video course added a visual and interactive element to your learning process, complementing the written information.

As you wrap up this book, keep in mind that the HVAC industry is always changing. Industry standards change, new technologies appear, and there are always new frontiers to explore. To further hone your talents, we urge you to maintain your curiosity, pursue lifelong learning, and think about enlisting in advanced courses or looking for mentoring.

Acquiring HVAC skill doesn't have to be an isolated path. Interacting with the larger HVAC community via online and physical means may provide insightful information, helpful hints, and a feeling of community. To continue learning and staying up to date on the most recent advancements, keep taking part in forums, going to industry events, and making connections with other fans and experts.

With this guide's information at your disposal, go ahead with the understanding that every setback is a chance for growth and every fix represents a step closer to mastery. Your path is distinct, and the knowledge and abilities you've gained here will serve as your guide whether you're taking on do-it-yourself tasks or thinking about a career in HVAC.

We appreciate your participation in our HVAC journey. I hope all of your HVAC (heating, ventilation, and air conditioning) pursuits bring you comfort, prosperity, and a never-ending desire for learning about and mastering the invisible symphony that creates the spaces we call home. I hope your voyage with HVAC is safe!

Appendix

Glossary of HVAC Terms

The inside part of a heating and cooling system that circulates air via ductwork is called an air handler.

British Thermal Unit (BTU): A heat energy measuring unit. It is the quantity of energy required to increase one pound of water's temperature by one degree Fahrenheit.

Condenser Coil: Part of an air conditioner or heat pump's outside unit where heat is released into the atmosphere via refrigerant.

A system of pipes or tubes called ductwork is used to move air throughout a structure.

The interior unit's evaporator coil helps with cooling by absorbing heat from the surrounding air.

An apparatus known as a heat exchanger is used to transfer heat without direct contact between two fluids, such as air and refrigerant.

HVAC, or heating, ventilation, and air conditioning, is the technology used to provide a comfortable interior atmosphere.

The effectiveness of an air conditioner or heat pump unit during the cooling season is measured by its SEER (Seasonal Energy effectiveness Ratio).

A thermostat is a device that modifies the flow of hot or cold air to regulate the temperature of a heating or cooling system.

Variable Refrigerant Flow, or VRF, is a kind of HVAC system that gives various areas of a building variable amounts of refrigerant flow.

Useful Websites and References

ASHRAE (American Society of Heating, Refrigerating and Air-Conditioning Engineers):

Website: https://www.ashrae.org/
A worldwide community that uses sustainable building technologies to improve human well-being.

ENERGY STAR:

Website: https://www.energystar.gov/
A program by the U.S. Environmental Protection Agency and the Department of Energy promoting energy efficiency.

HVAC School:

Website: https://www.hvacschool.com/
An online platform offering HVAC courses, podcasts, and a community for learning and discussion.

Contracting Business:

Website: https://www.contractingbusiness.com/
A resource for HVAC professionals providing industry news, trends, and business insights.

The NEWS - Air Conditioning, Heating & Refrigeration News:

Website: <u>https://www.achrnews.com/</u>

A publication covering the latest news and trends in the HVAC industry.

Thank you

I wanted to express my heartfelt gratitude for choosing to read my writing.

Your thought and thoughtfulness are much valued, and your support means the world to me.

I hope you enjoyed and learned from the book.

Thank you once again for reading and I look forward to sharing more with you in the future.

Best wishes,

Brandon Denney

www.ingramcontent.com/pod-product-compliance
Lightning Source LLC
Chambersburg PA
CBHW080955290526
45795CB00009B/2960